The Urban Book Series

The Urban Book Series is a resource for urban studies and geography research worldwide. It provides a unique and innovative resource for the latest developments in the field, nurturing a comprehensive and encompassing publication venue for urban studies, urban geography, planning and regional development.

The series publishes peer-reviewed volumes related to urbanization, sustainability, urban environments, sustainable urbanism, governance, globalization, urban and sustainable development, spatial and area studies, urban management, transport systems, urban infrastructure, urban dynamics, green cities and urban landscapes. It also invites research which documents urbanization processes and urban dynamics on a national, regional and local level, welcoming case studies, as well as comparative and applied research.

The series will appeal to urbanists, geographers, planners, engineers, architects, policy makers, and to all of those interested in a wide-ranging overview of contemporary urban studies and innovations in the field. It accepts monographs, edited volumes and textbooks.

Indexed by Scopus.

Nataliya Pankratova · Hennadii Haiko · Illia Savchenko

Modeling the Underground Infrastructure of Urban Environments

A Systematic Approach

Springer

Nataliya Pankratova
Institute for Applied System Analysis
Igor Sikorsky Kyiv Polytechnic Institute
Kyiv, Ukraine

Hennadii Haiko
Ins.for Energy Saving and Energy
Igor Sikorsky Kyiv Polytechnic Institute
Kyiv, Ukraine

Illia Savchenko
Institute for Applied System Analysis
Igor Sikorsky Kyiv Polytechnic Institute
Kyiv, Ukraine

ISSN 2365-757X ISSN 2365-7588 (electronic)
The Urban Book Series
ISBN 978-3-031-47521-4 ISBN 978-3-031-47522-1 (eBook)
https://doi.org/10.1007/978-3-031-47522-1

This Springer imprint is published by the registered company Springer Nature Switzerland AG
The registered company address is: Gewerbestrasse 11, 6330 Cham, Switzerland

Introduction

The growth of large cities is a manifestation of steady historical tendencies, and it leads not only to the constant expansion of metropolises but also to the significant complication of their functional and spatial organization. Furthermore, in many cases, the traditional potential of city growth "upwards and broadwise" is exhausted, and complex mastering of urban georesources has started. Dealing with a number of acute problems (territorial, logistical, power supply, ecological ones, etc.) related to the intensive metropolis growth can be successfully achieved by constructive development of urban underground space. Regulating urban development with the goal of increasing comfort, ecological standards, and life safety in the constantly growing metropolises is one of the most urgent though insufficiently studied and complex global problems.

Pivotal changes that happened in the last decades in the life of big cities require scientific comprehension of new reality and most likely prospects of cities' further growth. Urban underground development, which is an integral part of modern metropolises, has already gone beyond the scope of separate local objects and has become a systemic existential factor of large cities. The foresight of upcoming changes, the construction policy, and the city planning for metropolises should be based on a reliable scientific and methodological foundation which provides the harmonic evolution of surface and underground urban development as a whole.

The modern approach to metropolis planning is based on the system concept of sustainable city development. It should satisfy the urgent societal demands without losses and harm to future generations. The important aspect of sustainable city development is the potential for timely reaction to changes in structural, functional, and natural environments in order to minimize technogenic and ecological risks. This concept impacts the scope and strategy for many engineering projects and reckons for the shift from traditional addressing of local problems to the contemplation of a project as a part of large natural, technical, and social systems.

The urban underground space is capable of providing three-dimensional freedom of movement for people, materials, water, and energy resources, particularly to the hard-to-reach places in densely built-up districts of a city. Millions of people nowadays rely on underground communications that ensure the safety and comfort for city dwellers. In the sustainable development concept, the urban underground development has special significance, since an efficiently planned underground infrastructure increases the life quality and ecological safety to a greater extent compared to a similarly functioning surface system. Due to the global trends in the increase of military and terrorist threats caused by the consequences of the Russian invasion of Ukraine, the potential of the underground space must be used to the fullest extent for the protection of critical infrastructure and the civilian population. This can be achieved by the transfer of the functions of the most critical and vulnerable surface objects to underground structures.

The complexity of mastering underground space of large cities is caused not only by the need to substantiate the structural, functional, and spatial decisions but also by the variable properties of the geological environment containing the underground facilities, creating additional impact factors and risk elements. A substantial increase in underground construction risks is observed in the area of influence of water bodies, demanding more efficient construction site (tunnel track) planning. Development of the foresight and risk management culture should be an important part of design thinking for a modern underground city planner, along with the creation of a universal systemic tool set for underground space planning.

Despite the active underground construction in many of the world's metropolises (examples worth noting are Helsinki, London, Montreal, Osaka, Beijing, Singapore, Tokyo, and Shanghai), the underground development potential is still insufficiently tapped. Projects of underground structures mostly remain separate, local, and loosely connected both with each other and with the surface objects; the construction itself is conducted without the strategic plan of developing an "underground city" and without the scientific analysis of alternatives, serving as a palliative decision of complex problems. The limited extent of utilizing underground space is caused not only by the investment climate issues but also due to the weakness of territorial development policy, insufficient level of legislative and regulatory basis for property issues of geoconstruction objects, the lack of an agreed concept and strategic master plan of urban underground development, and, as a result, the absence of attractive well-grounded proposals for global investment groups. Such typical problems are encountered, for example, in the development of the underground space of the Kyiv City, where underground construction in the medium term should increase significantly (according to the General Development Plan of the Kyiv City).

The authors set their task to highlight the advantages of the systemic approach to mastering underground metropolis space and to present the possibilities opened up by the system analysis methods for risk assessment and prediction of the development scenarios for urban underground construction as a system of alternative project configurations, that may become the basis for creating the strategic master plans of "underground cities" and favor the mechanism of assessing socio-economic efficiency of mastering underground city space of metropolises.

The Chap. 1 presents the historical excursion of constructive mastering of underground space, the analysis of several modern trends in global urban underground development, and the principles and prospects of its sustainable development.

The Chap. 2 deals with the main principles of the system approach to planning underground space of metropolises. The natural-technical system "urban underground development—geological environment" is proposed, viewing the interaction of natural and technogenic factors with structural-functional factors of mastering underground space. The technique for typing and districting geological environment of a metropolis is developed, which evaluates the territory for favorability of underground construction.

The Chap. 3 overviews a powerful applied system analysis tool—the modified morphological analysis method which is an important asset for efficient planning of mastering underground city space.

The Chap. 4 shows a number of practical applications of the modified morphological analysis method as a tool for solving various urban problems, including risk assessment in transport infrastructure of a city, and the evaluation and justification of constructing different types of underground objects on a selected site in accordance with its structural, functional and geological factors.

The Chap. 5 considers the application of another powerful system approach—the combination of BOCR (benefits, opportunities, costs, and risks) analysis, and analytical hierarchy process method for practical tasks of urban development.

The Chap. 6 describes the application of the methodologies of foresight and cognitive modeling for simulating complex underground objects, lays down the theoretical foundation of foresight and cognitive modeling methodologies, and shows their application on several example scenarios for underground construction.

Thus, readers are offered one of the few works that considers system models and planning solutions of surface and underground urban development as a whole from the standpoint of sustainable development of large cities, providing a complex process of planning decision-making using the methods and tools of applied system analysis.

The authors express their gratitude to Viktor Kravets for his valuable contribution to the development of the concept of the natural-technical system "Geourbanism—Natural Environment" (Sects. 2.1 and 4.3), to Nadiia Nedashkivska for her expertise in the modified BOCR method and its facilitation for the studied problems (Chap. 5), and to Volodymyr Pankratov for his important role in cognitive modeling research (Chap. 6).

Contents

Chapter 1
Historical Excursion and Modern Trends of Urban Underground Development

Abstract The historical stages of mastering the underground space by humans are presented, and the significance of the geoconstructive mastering of earth is justified. The first steps of urban underground construction at the Industrial Age were related to the invention of the tunneling shield, and the shielding method of excavating tunnels in the ground. The first attempts at a system approach to planning a network of underground structures during the creation of the London metro are analyzed. The important achievements of urban underground construction within the latest half century and the trends of future developments of underground structures are discussed. Concept projects of the large-scale underground complexes are presented, planned for construction in the world's metropolises as peculiar "underground cities". The authors' concept of sustainable development of large cities as a synthesis of surface and underground urban development is presented. The most urgent tasks for legislative, legal, educational, and scientific management spheres are justified by the goal of intensification of the underground space mastering in accordance with the demands of the modern large cities.

1.1 Historical Excursion in Geoconstructive Mastering of the Underground Space

Mastering underground space by primitive humans should be viewed as a cultural phenomenon of their interaction with the natural geological environment, which has started in the Paleolithic and has been going on in parallel with the human evolution. Anthropologists mark the "cave period" of human existence, when the main shelter for people (in the terrains with an alpine or a hilly landscape) were natural caves, where humans were seeking protection from unfavorable weather conditions, savage animals, and aggressive neighbors. The oldest Peking Man cave settlement was found in northern China (Zhoukoudian Cave), which was exploited by primitives ca. 250 thousand of years ago. Kapova cave, one of the most famous due to the rare cave paintings, was colonized by humans in Ural foothills ca. 200 thousand years ago. Homo Sapiens, which emerged on the territory of the African continent, significantly

N. Pankratova et al., *Modeling the Underground Infrastructure of Urban Environments*, The Urban Book Series, https://doi.org/10.1007/978-3-031-47522-1_1

expanded the scope of utilizing natural underground objects. Adaptation of natural caves to their needs (expansion, deepening, interconnection) should be considered as the first intentional human activity of technogenic mastering of underground space, followed by the archaic activity of mineral extraction. The initial experience and skills of reconstructing existing natural underground objects were used for creating man-made cave settlements and sacred formations, and significantly improved with the emergence of metal tools for breaking rocks (starting with Eneolithic) (Haiko and Biletsky 2015).

A large number of cave cities were constructed at various times in the Mediterranean region, the Caucasus, the Crimea, the Near and Middle East, and India. As an example, we'll consider the cave city complex in Cappadocia (central Turkey), the construction of which started in the Hittite times (II millennium BC), and achieved its peak of growth in the early Middle Ages, as a settlement of Christian communities (Shylin 2005; Spiro 1989). The total number of discovered cave cities in Cappadocia reaches thirty-six. A special geological structure of the region was caused by the ancient volcanic activity, numerous magma outpourings, covered by massive volcanic tuff, with further processes of weathering and erosion. It provided favorable conditions for construction, as volcanic tuffs can be easily mined by metal tools (pickaxes, chisel and hammer, wedges, etc.), simultaneously keeping their stable properties, which even increase under the influence of air. The conducted experiments show that a worker equipped with a metal tool could excavate up to 100 m^3 of rock monthly.

The most explored is the ancient city of Derinkuyu (deep well), that was discovered in 1963, and in 2 years, the first tourist track was created there. The city takes up an area of 4 $\times$ 4 km and has nearly 20 underground levels (the deepest shaft reaches 120 m), of which only 8 are examined (up to the depth of 55 m). To provide ventilation, drainage, and water necessities, 52 vertical shafts were constructed, that reach groundwater (at 85 m depth). Ventilation shafts were connected in the lower part with the fire chambers where a bonfire was constantly maintained, providing a flow of warm air upwards, and absorption of air streams through adjacent excavations that reached surface. The city has nearly 2 thousand rooms (churches, living chambers, animal lodgings, water reservoirs, storage rooms, wine cellars, countless passageways, etc.), and has up to 600 surface exits. The area of the largest underground chamber reaches 300 m^2. Some assessments estimate the total population living simultaneously in Derinkuyu to be up to 10 thousand persons. The large population, the amount and functionality of the underground facilities indicate efficient planning decisions of mastering underground space and systemic thought of its creators (although the settlement was founded as separate local caves, the rise of their numbers bred the idea of creating an essentially new interconnected "cave city").

It should be noted that the sacral, religious factor for a long time was a potent stimulus for broad mastering of the underground space. Among the sacred underground construction, several types of structures can be highlighted: tombs (burial chambers or crypts with ceremonial chapels); temples (underground constructions of various forms with wide areas, chambers of large cross section with an altar and places for believers); monasteries (lodgings for cult acolytes and believers near a temple, or separately); and complexes of sacred buildings that combine the listed objects in a single underground space.

As one of the most impressive tombs that lived up to our time, Ancient Egyptian pyramids should be noted, as they contained several immersed burial chambers with sarcophagi, and a system of vertical, horizontal, and sloped tunnels, faced with stone slabs (Shylin 2005; Lehne 2008). Pyramid of Djoser (2700 BC) was a family crypt with a central shaft that had a cross-section of 7 × 7 m and reached the depth of 27,5 m; the surface construction reached 60 m in height. The pharaoh Djoser's burial chamber copied his palace in Memphis. A curious feature of this sacred complex is its unity of underground and surface sections.

Territories relatively poor in rock materials were home to magnificent burial constructions erected in another manner. A peculiar example is the graves (kurgans) in Ukrainian steppes that had an immersed burial chamber (sometimes faced with stone slabs), and a huge mound erected from stones and soil. A notable sample is the Royal Kurgan near Kerch, erected at the time of the Bosporan Kingdom in the fourth century BC. Its height exceeds 17 m, outline—260 m; a special tunnel—dromos (length 36 m, height 7 m), faced with stone slabs, leads to the underground chamber of an arrowhead cross-section (Haiko et al. 2009). Hundreds of kurgans by different peoples, including the Scythians and the Slavs, are present in the Northern Black Sea region, signaling the inheritance of the tradition of burying noble persons in kurgans. An interesting sacred monument of Ukraine is its Stone Grave near Melitopol, which served in millennia as an altar for religious ceremonies, and contains several thousands of unique ancient rock carvings—petroglyphs (Rudynskyi 1961).

Various religious beliefs of different peoples found their representations in architectural embodiments such as underground temples, which were often constructed in mountain ranges. The ancient Egyptian majestic temple of Ramesses II in Abu Simbel was carved out in a massif of pink sandstone in the twelfth century BC (chamber size: 17.5 m length, 16 m width, 10 m height; chamber ceiling supported by 8 bearing pillars in the form of columns; the sanctuary extends 60 m deep into rock). A complex of cave temples and monasteries in Petra (Jordan) was constructed ca. third century BC and at the beginning of the Common Era became a religious center of the Near East. Middle Age India is also famous for impressive construction of cave temples. Magnificent temple complexes were created directly from large rocks, with their surface skillfully processed, and vaulted chambers cut out inside. Among the largest temple complexes are Ajanta (29 cave structures) and Ellora (34 cave temples) in Central India. A similar practice, although for Christian temples, was employed for rock temples in the Middle Age Ethiopia.

The "underground cities" of some sorts were the numerous catacombs of large cities that emerged as cavities (excavations) after mining building stone. With time the free underground space was transformed into sacral complexes used as cave temples, monasteries, and burial grounds. One of the most famous are the Roman catacombs which became a hiding place for the first Christians. The chambers of underground quarries were reconstructed into church basilicas, chapels, and crypts, and laid the foundation of the Christian temple architecture, that later emerged out of "hiding" to the surface in mostly untouched architectural planning appearance of the first churches (Fink 1997). The Catacombs of Callixtus are the lengthiest in Rome: the total length of tunnels reaches nearly 20 km, and the depth of the lower floor is 20 m. The mastering of the underground space of catacombs is one of the first examples of the "postuseful" exploitation of underground structures, when they get a new functional purpose in time.

Listing cave temples and monasteries, we should mention also Kyiv-Pechersk Lavra, as its very name talks about the cave origin of the monastery. The first monks' caves were constructed in the eleventh century (excavations were made in loess soil). Later a system of underground tunnels and chapels (churches) was made under the monastery territory. Their depth varies from 5 to 20 m, the tunnel heights 2–2.5 m, and width 1.2–1.5 m. The complex is a ringed maze totaling 288 m in length with two exits. Walls in some sections are faced with bricks. Both near and far caves contain 3 churches each, and a number of cave burial sites (Haiko et al. 2009).

Aside from the underground sacral buildings, settlements, and military fortifications, the underground space was utilized long before the Common Era as an efficient communication and supply route (engineering communications), by making hydrotechnical, transport, and pedestrian tunnels. The oldest underwater tunnel, which is known from antique sources, was constructed under the Euphrates River and connected two parts of Babylon (the Marduk temple and the royal palace) (Passek 1933). Diodorus Siculus in "Bibliotheca historica" describes the construction process, and using the tunnel by Semiramis. He notes that firstly a dam was constructed that rerouted the Euphrates waters from the tunnel construction site. Its construction was made using the open (trench) method. A tunnel was fortified with strong bricks and covered from the inside and outside with a thick layer of bitumen. Then the immersed structure was covered by soil, and the river was let in the old bed. The tunnel length was 929 m, width was 4.6 m, and height was 3.7 m. Some archeological findings date the tunnel at the end of third millennium BC (another version puts it at Nebuchadnezzar I times, i.e., twelfth century BC).

One of the most famous hydrotechnical tunnels in the world is the Siloam tunnel in Jerusalem that transferred the water from Gihon Spring to the Siloam reservoir, providing the city with water. Erected under the City of David during the reign of Hezekiah of Judah (the end of eighth-the start of seventh century BC), the tunnel is mentioned in two Kings. The length of the construction is 533 m, height—3 m, and depth in the middle section—50 m. The Siloam tunnel got its wide fame due to a fragment on one of its walls that describes the tunnel construction process.

According to the Siloam inscription, the tunnel was excavated by two groups, each starting from one of the tunnel's exits, and meeting in the middle, which attests to the high skill of ancient geoconstructors (the moment of the meeting itself was described in detail by an eyewitness) (Frumkin and Shimron 2006). The Siloam inscription is the first documental source from the history of urban underground construction. Nowadays, the Siloam tunnel is included in the tourist track "City of David in Jerusalem".

No less famous ancient hydrotechnical construction is the sewage system of Ancient Rome, particularly its large sewage canal Cloaca Maxima. It was built in the sixth century BC in a system of drainage tunnels between the Palatine Hill and the

Capitol, becoming in time the largest sewage collector of Ancient Rome (Hopkins 2007). The tunnel length is 320 m and its cross-section—3 × 4 m. The tunnel's facing is made using hewn tuff and travertine with inclusions of rubble and brick. The tunnel's portal was fixed by vaulted masonry from tuff wedge blocks placed in three rows. Sewage was drained to the tunnel through vertical holes and through the side tunnels with different intersections. Sewage was then poured into the Tiber River. The Rome's sewage system provided great comfort and significantly increased sanitary conditions in the city. Only a few ancient cities of the world outpaced Rome in this civilizational acquisition (Mohenjo-Daro, Babylon, Linzi); however, it was Rome that became an example for implementation of these engineering communications into European city space.

A number of curious tellings about hydrotechnical tunnels were found in antique sources. "The Father of History", Herodotus attributed to the seven wonders of the world the tunnel on the Samos Island in the Aegean Sea, that was constructed in the sixth century BC by the Ancient Greek engineer and geometer Eupalinos of Megara (the tunnel thus is also known as the Tunnel of Eupalinos). The 1 km long tunnel was excavated in sturdy limestone and was designed for water supply of the island polis. A peculiar feature of the tunnel is its water supply canal in the bottom part 9 m deep and almost 1 m wide. The tunnel was excavated from both ends. The Samos tunnel was rediscovered in the 1880s and is exhibited as a tourist attraction (Olson 2012).

Over 300 km of hydrotechnical tunnels were excavated for water supply and melioration of the Roman Empire cities, some of which were connected with the aqueducts. Their perfect construction became a "calling card" of the Roman engineers. Pliny describes the underground construction of the tunnel in the Fucine Lake region, with a length of over 5,5 km and 6 m height. To create the tunnel, 40 vertical shafts up to 120 m deep were excavated. During 11 years of the tunnel's construction, up to 30 thousand workers were involved. The tunnel started operating half a century before the Common Era to dry the Fucine Lake which was waterlogging the farmlands and a source of mass malaria cases. Suetonius in "The Lives of the Twelve Caesars" attributed the fame of constructing the Fucine Tunnel to the emperors Claudius and Hadrianus, noting that the financing of the project was gathered with the purpose of handing the drained lands to the investors.

A similar grandiose object was erected by the Romans in the middle of the first century AD near the seashores of modern Albania. A 5.6 km long and 2.7 × 5.8 m in cross-section (narrow and high form of a tunnel was more stable) tunnel was excavated for water supply of the colonial city. The construction was made simultaneously with several excavations from intermediate shafts. Over 2 km of the excavation was cut through strong lava rock. The construction involved 20–25 thousand workers (Lykhin 2003).

All of the aforementioned constructions were created exclusively with the handwork of slaves using unproductive hand tools, corresponding to the productive forces level of that time. Strong mountain rocks were cut using the miners' adopted "incendiary method": a rock was heated in a shaft with a large bonfire, then drenched in water, forming cracks after an abrupt change of temperature. Using chisels and hammers, the rock was further crumbled to prolong the shaft.

The fall of the Roman Empire and the slave society made the construction of such grand structures organizationally impossible, so it was paused until the Industrial Revolution and the emergence of the new technological capacities of humanity. Some large-scale objects (e.g., the above-mentioned Derinkuyu City) were erected in the Middle Ages for a very long time (several centuries) and were stimulated by religious factors. Even underground fortification structures that performed the life support and defense functions of the fortified cities were significantly smaller in scope than those of the ancient world.

A significant impact on the scale of the underground construction was made by utilizing explosive materials for wrecking stone and making excavations. The first documented case of using gunpowder for creating a mining excavation shaft was dated 1627 (Banská Štiavnica, Slovakia) (Haiko and Biletsky 2015). Since that time, the improvement of drilling and blasting works has started in the directions of designing machinery and equipment for making blast holes, increasing efficiency and safety of explosive materials and detonation means. Significant technological advances of the nineteenth century benefited abrupt increase in tunnel construction, with the speed of excavation reaching 80 m a month and more (Lykhin 2003). Navigable tunnels and canals also became widespread. Some cities were making their systems of underground water passages for transporting cargo and goods by boats to storages and shops (in Ukraine this was a case for Kamianets-Podilskyi city).

Urban underground construction for a long time was dominated by the open method (the underground structures were erected in open trenches or pits, and then covered by soil). Fundamentally new capabilities were provided by the revolutionary invention of Marc Brunel who created a tunneling shield for underground excavation in weak rocks (soils). The inventor made the following conclusion: in soft soils, the casing in a tunnel should be constructed right after the advance of the excavation, and the excavated space at the perimeter of the tunnel should be protected from collapse by a cylindrical construction (shield). In 1818 Brunel patented the tunneling shield which was moved by screw jacks that leaned onto the tunnel casing built before the shield (Passek 1933). The operating body for excavation was of rotary type and was actuated by manpower. In 1825, using his invention, Brunel started the construction of the tunnel under the Thames River, though it was (due to various reasons) finished only in 1843.

This was the first underground tunnel in the world constructed in weak soils. The tunnel under the Thames was 11 m wide and 6 m high. At first, it was exploited as a pedestrian tunnel, and starting in 1865, the passage of trains in it has begun. Thus, it was London City that began the tradition of large-scale underground construction in the new times. No wonder it was London where the first project of the urban underground railway was initiated and implemented, known under the name of "metropolitan" (from the name of the first line "Metropolitan Railway", i.e., "capital railway", that connected Paddington and Kings Cross stations with the Farringdon station). An interesting fact is that the idea of separating pedestrian and transport flows at different levels was proposed long ago by Leonardo da Vinci.

Only the Industrial Age characterized by mass influx of workforce into cities set up an issue of the large-scale mastering of underground space of metropolises, since its economic benefits already became evident in the second half of the nineteenth century. A bursting growth of English railways that had radial scheme and converged from different directions into London caused the construction of six dead-end suburban railway stations. Each day tens of thousands of people arrived to work in London from surrounding towns and villages; the transit passenger flow increased considerably, and the capital's population exceeded 2 million. The horse-buses and private cabs could no longer satisfy the logistical needs of the city, traffic jams became an everyday phenomenon. Extending railways closer to the city was prohibited by technical norms (the danger of ruining buildings caused by dynamic loads). The London authorities started to consider an issue of constructing the urban underground railway at the start of the 1850s (the first propositions were made as early as the 1830s). In 1855 the construction of the underground railway called the "Metropolitan Railway" was affirmed by a parliamentary act. This cause inspired the London lawyer and social activist Charles Pearson who lobbied the allocation of funds to the implementation of the project in the municipal parliament. The construction (using the trench method) started in February 1860 under the supervision of the chief engineer John Fowler. The first line of the London Metropolitan (using steam traction) opened o January 10, 1863 (Ackroyd 2015).

In the very first year of the metropolitan's existence, ca. 10 million passengers were carried, and this figure increased each year. The construction of new shallow laying lines started. The main shortcomings of the open (trench) construction method were the traffic terminations at the streets where construction was conducted, and the demolishing of the surface infrastructure (individual houses at the "critical" sections in particular). Additionally, shallow laying lines paralleled the city streets, which were not always the shortest and rational route.

In 1890 in London, the first deep-laying metropolitan line was founded, constructed with the help of the tunneling shield. This line was one of the first to be equipped with electrical locomotives. Deep laying of stations and tunnels made passengers' descent difficult and uncomfortable by long stairways, leading to the first use of escalators, which in time became ubiquitous and almost a symbol of mastery of underground space. The project's implementation was made under the supervision of the British engineer James Henry Greathead. Deep laying of the tunnels had very little impact on the surface buildings, allowing to set the metropolitan lines under residential blocks in the desired directions.

The four lines of the London metropolitan (including the famous Central one) were constructed with the funds of the American financier Charles Tyson Yerkes who, using the system approach, tried at the start of the twentieth century to create a single underground London network at the place of separate tunnels and local underground facilities. Due to the tunneling shield structure, the tunnels had an almost round cross-section, producing the well-known name of the London metropolitan—"The Tube".

The success of the transport mastering of London's underground space was so evident, and the solution of a number of acute problems of a large city so illustrious that the metropolitan construction, despite the high capital expenditure, became widespread, and, among other, a distinguishing feature for cities and countries. In

1875 the first attempt at building metropolitan in Istanbul was made (the constructed "Tünel" 573 m long was operating like a funicular). In 1897 in Boston, the first metropolitan in the USA started working, in 1904—in New York, in 1907—in Philadelphia (surface metro lines were constructed in the USA since 1885). In 1896, the first metropolitan lines were constructed in Budapest and Glasgow, Budapest metropolitan being the first in continental Europe. The following chronology of opening metro lines looks like 1898—Wien, 1900—Paris, 1902—Berlin, 1912—Hamburg, 1913—Buenos Aires (the first in South America), 1919—Madrid, 1924—Barcelona, 1927—Tokyo (the first in Asia), 1935—Moscow, 1950—Stockholm, …, 1960—Kyiv, etc.

The start of the twentieth century marks the new architectural directions of the growth of large cities, related to the systemic multifunctional use of the underground space, with metropolitan remaining its pivotal development core. Despite the utopian nature of some projects that were "ahead of time", these concepts set the prospective trends of uniting the separate underground structures into a single underground complex, interconnected with the surface buildings. In 1933, an international organization for studying the urban underground development problems, GECUS, was established, with its central principle being "There's no urbanism without underground urbanism". In the 1930s, a model of "vertical segregation of city functions" is formed.

From the 1930s to the 1960s, the defense objects become an important constituent of underground construction, particularly the government and military command centers, and the civil defense shelters (Diakov et al. 2018). Among the most peculiar constructions of sort, we can name the German command quarters Maybach near Zossen city (seven underground levels, constructed in 1936–1940), the London bunker Berlington (total area over 1000 m^2, personnel amount—3000 persons, 1950), command center of aerospace defense of the USA under Cheyenne Mountain, Colorado (underground city for the case of nuclear attack, constructed in the 1960s), underground shelter city in Beijing (expanding over 30 km long, it could at the beginning of the 1970s house up to 40% of capital population), underground complex for sheltering submarines in the Ukrainian city of Balaklava (Crimea, "object #825 GTS", 1957–1961), underground bunker under the presidential palace in Baghdad (in 2003, it withstood the direct hit by Tomahawk missiles), etc. (Lysikov and Kapliukhin 2005; Bryk et al. 2012; Egorov and Aksyonov 1996). During World War II, the metropolitan became the most reliable shelter against hostile barrages and bombardments, particularly in London, Berlin, and Moscow. Underground structures of civil defense became the most widespread objects of urban underground construction even in the post-war period, though due to the increase in nuclear weapon power, the shallow laying structures lost their potential of reliable protection of residents, so their mass construction was canceled, further favored by the end of the "Cold war" at the beginning of the 1990s. The Russian aggression against Ukraine that started in 2014 and grew to the scale of the full-fledged war in 2022, reinstated the urgency of protecting population and critical infrastructure in underground space, renewing the demand of using underground structures as double-purpose objects.

Summarizing the long-term experience of using the underground space by mankind,[1] it can be noted that at the first stage people's dwelling underground was caused by the vital necessity to seek protection from unfavorable weather conditions, natural phenomena, and enemies, as well as under the influence of sacral (religious) factors that led to creation of countless underground temples, monasteries, burial chambers, religious settlements. The industrial age and the massive influx of rural population into industrial cities, their bursting growth set up a problem of urban underground development, i.e., the mastering of urban underground space of metropolises, particularly for solving logistical issues. The development of geoconstructive technology and machinery made it possible to use underground space as an efficient transport and communicative resource, and its protective functions largely provided the sheltering of urban population at periods of war.

1.2 Modern Achievements and Future Trends of Mastering Metropolises' Underground Space

In the last quarter of the twentieth century and at the beginning of the twenty-first century, the search for optimal methods of mastering underground space was predetermined by the following factors: facilitation of historical downtown areas, limitations on density and height of buildings near historical and architectural monuments, shortage of free land sites for construction, development of service industries in the most crowded places (trade, food, cultural events), development, and arrangement of the transportation field including temporary and regular parking of cars, transferring urban engineering communications to underground space, saving fuel and energy resources, forming potential civil defense objects in case of a war. Since 2022, the issue of using underground space as a civil defense and critical infrastructure protection area, and viewing underground complexes as double-purpose objects, renewed its urgency.

Global interest in mastering underground space is largely fueled by the advantages of underground constructions which include lowering expenses on heating and cooling of premises, lowering operational expenses compared to the surface structures, and significantly decreasing the impact of climate conditions and technogenic factors. The capability of the ground volume to reliably safeguard people against the hazardous external impacts allowed to widely use the underground structures as protective shelters from mass destruction weapons, military strikes, natural disasters, and technogenic catastrophes (State Construction Norms 1998; Rudniak 2003; Golubev 2005).

[1] Another direction of wide mastering of the underground space was mineral mining; furthermore, the mining technologies sometimes found their use in geoconstruction. However, the topic of mineral resources excavation lies beyond the scope of this book.

The most noticeable problem of large cities becomes the shortage of territory, caused by discrepancy between their structure and the architectural planning decisions, and the exorbitant rise of population and surface transport which wasn't planned or even predicted in times of the emergence and primary formation of the historical cities. The world urbanization rates bear an expressive statistic. In 1800, the number of urban dwellers on the planet equated to 5%, in 1965—32%, and at the beginning of the twenty-first century this indicator equated to 91% in Germany, 89% in Israel, 85% in the USA, 81% in Japan, 76% in Great Britain, and 67% in Ukraine. Urbanization in the twenty-first century is characterized by the growth of megacities and the spread of enormous urbanized conglomerations. Over 20 million people (including suburbs) reside in Shanghai, Tokyo, Delhi, and Mexico City; 10–20 million reside in Karachi, Lagos, Istanbul, Guangzhou, Mumbai, Beijing, Dhaka, Moscow, Cairo, São Paulo, Osaka, Seoul, Jakarta, and New York (near 10 metropolises are nearing this category, and 64 cities have more than 4 million residents). The concentration of metropolises, their suburbs, and satellite cities create agglomerations that fundamentally change the natural society state and breed the specific, unseen before-living manners, and generate new societal and philosophical challenges (Prepotenska 2014). The density of population here varies from 3 to 7 thousand persons per km^2. For example, a quarter of the USA population is concentrated along the Atlantic shore (Boston to Washington). Examples of the largest conglomerates include Tokyo-Osaka (66 million), Chicago-Pittsburgh (35 million), London-Liverpool (30 million), and San Diego-San Francisco (20 million).

Relatable processes, typical for many countries in the world, touched also Ukraine, which at the start of the twentieth century was recognized as a "granary of Europe" and an exemplar of the rural existence and outlook. In 2017, the number of urban residents comprised 71%, 459 settlements had a city status, 5 of which had over 1 million population: Kyiv, Kharkiv, Dnipro, Odesa, and Donetsk (before occupation); 4 more had 0,5–1 million population (Zaporizhzhia, Lviv, Kryvyi Rih, Mykolayiv); and in 37 cities resided 100–500 thousand people. Kyiv population including suburbs now comprises over 4 million residents. Half of the country's urban dwellers live in agglomerations, which amount to nearly 20 in Ukraine, the largest of them being the Kyiv, Donetsk, Kharkiv, Dnipro, Lviv, and Odesa ones (totaling over 12 million residents). Therefore, Ukraine's urban development corresponds to the common global trends and meets the same challenges, which can be largely solved by urban underground construction.

For most of the metropolises, the potential for traditional development "upward and broadwise" is nearly exhausted. Urban underground development becomes an important growth direction. As a researcher and chronicler of London, the novelist Peter Ackroyd noted, "the prospect of the metropolis is to reach deeper and deeper underground" (Ackroyd 2015). The Japanese city building convention is "as high as the city grows upward, that much it should descend downward". The underground areas of some modern metropolises with the most developed underground parts (Helsinki, London, Montreal, Tokyo, Osaka, Beijing, Shanghai, and Singapore) comprise 20–25% of their respective surface areas. Most of the cities with constructed metropolitan networks have this number at 5–15%, indicating a significant potential for growth of urban underground construction. At a certain stage of a metropolis' growth, a shift is made from design and construction of separate underground structures to multi-layered complexes which partly replace surface city functions for residents (so-called "underground twin cities").

Modern city building concepts envision the overall distribution of all urban underground structures onto four layers of depth that sometimes delegate their functions to adjacent layers:

- layer one (at maximum depth)—communications that operate without constant human presence;
- layer two—industrial and power enterprises with constant presence of a limited number of qualified personnel;
- layer three—transport tunnels, car garages and parking lots, storage space and auxiliary facilities, service communications with short-term use by a large number of people;
- layer four (shallow laying)—pedestrian zones and adjacent trade, sport and entertainment centers, hotels, social and administrative institutions, that operate continuously and provide long-term presence of a large number of people.

Let's take some examples of large modern underground complexes (Kelemen and Vayda 1985; Kaufman and Lysikov 2009; Sterling et al. 2012).

The underground city in Montreal (The Underground Siti RESO), also called "the inner city", is one of the largest underground complexes in the world. Its total length of underground communications reaches 32 km, and it bears 40 large objects in the total area of 12 km^2. Everything needed for comfortable living is concentrated here: hotels, trade centers, universities, museums, sport facilities, banks, business centers, metro stations, railway and bus stations, and other entertainment and business establishments. Adjacent surface buildings are connected with underground facilities by passages (there exist over 120 surface entrances) that provide movement of large crowds to the underground city space directly from buildings, bypassing surfaces. It guarantees convenient navigation of underground space and transfers to the required city regions by underground routes, which is especially rational given the harsh Canadian climate. The underground city is daily visited by over 500 thousand denizens.

Underground Tokyo is a set of several complexes which are either operating or under construction, the deepest of which reaches 16 levels underground, and the peculiar features of mastering the underground space in the city are the largest pace and scale of underground construction (only in recent years has this record passed to Beijing). Yearly throughput in Tokyo metro is the second biggest in the world (3,62 billion passengers), with its dynamics explained not only by the constant introduction of new lines and stations (their amount reaches 291). Each Tokyo metro station is connected to a multistoried car and bicycle parking lot, and a pedestrian tunnel network that allows passage to all residential blocks served by the station. The union of stations and multifunctional underground complexes that comprise large trade and business blocks form the transport, social, and cultural underground centers of the main regions of Tokyo.

In the western part of Big Tokyo, an underground city Geotropolis is being constructed that utilizes the new concept of an underground residential environment, reliably protected from potential earthquakes, typhoons, and tsunamis. It could be the first time in modern history that permanent (long-term) being of residents is anticipated in a comfortable underground environment formed from the underground streets of residential and trade blocks, and express transport communications. The city is located in a strong layer of argillite 50 m deep from surface, providing favorable conditions for reliable support of underground constructions. To solve the psychological problem of long term being deep in a closed space, a set of cutting-edge engineering measures is used for imitating daily surfaces (including the appropriate lighting, moving "window scenery", sounds, scents, etc.). Planning underground space as a living environment sets new challenges for human psychological adaptation, that should combine engineering and socio-humanitarian techniques of urban underground development (Prepotenska 2014; Kelemen and Vayda 1985; Kartosiya 2015; Shemyakin 1995).

Another interesting part of Tokyo's underground world is its network of tunnels for prevention of floods during deluges or tsunamis. In the case of catastrophes, the water is expected to be rerouted to special underground large-size facilities (a system of underground bunkers 65 m high and 32 m wide), and then drained by 6 km long tunnels to the ocean, saving the city from flooding.

The capital of Finland, Helsinki, is probably the only city in the world which had its underground component created not as a sum of independent underground objects separately constructed at various times, but as an underground city designed by a single plan which was conceived in 1972 (Vähäaho 2014). Nowadays, it has more than 400 interconnected underground objects varying in scope and complexity with an overall area of nearly 9 million m^2: trade and entertainment centers, unique sports infrastructure objects (including swimming pools, stadiums, skate parks), concert halls, museums (including the new Center of modern arts "Amos Rex"), railway stations, underground parking lots, religious buildings, etc. Underground Helsinki has over 200 km of tunnels with large cross-sections (up to 21 m^2), containing water and heat supply pipes, sewer pipes, and a unique system of automated waste disposal lines that replaced garbage trucks and garbage bins from the city streets. Drinking water is supplied to Helsinki from a countryside lake by the largest tunnel in the world (120 km long, 3 m wide, and 5 m high).

The experience of creating Helsinki's underground twin city can be useful for many large cities where a massive underground construction is planned, however, each metropolis has its own distinct geological engineering, landscape, and hydrological features, unique goals and tasks of urban underground development, therefore a routine application of a specific construction project to another city will not be successful. On the other hand, a system planning approach demonstrated by Helsinki came out efficient and prospective. The largest cities of Japan (Tokyo, Osaka), China (Beijing, Shanghai), the USA (Minneapolis), Singapore, etc., are now following Helsinki's path.

The construction of a ramified transport infrastructure remains one of the key directions of mastering underground space of large cities, with the metro network complemented by networks of express car and railway tunnels. Permanent development of metropolitan led to high-intensity underground passenger transfers in many metropolises of the world. Modern London exhibits 270 stations and 402 km of transport tunnels; in Beijing, respectively, 391 and 637 km, in Shanghai—413 and 676 km, in New York—472 and 380 km, in Tokyo—291 and 304 km. Compared to the leaders, the Kyiv metro is relatively humble—52 stations and 69.6 km of tunnels; however, these indicators are similar to most of the European cities. Considering that each station is a complex of underground facilities (adjacent passages, stores, and food establishments), metro stations can be called the peculiar foci of "underground life" and a planning framework for an "underground city" development, which is used daily by millions of people. To perceive the scope of the underground passenger throughput, let's just present the data for the three metro leaders of the world: Beijing—3,66 billion passengers a year, Tokyo—3,62 billion, and Shanghai—3,4 billion.

Among the biggest modern projects for express train connection let's note the *Crossrail* in London, which when combined with metropolitan should solve the London transport issues according to the population growth forecast. Project funding already overcame 20 billion British pounds, 42 km of tunnels and 11 stations are already constructed. The trains that will cruise these two new lines each 2.5 min are able to carry 1100 passengers and connect the Heathrow airport and Kingston on the West, and Shenfield and Abbey Wood on the East.

No less importance do the urban automobile tunnels have. The first car tunnel in the world was constructed in 1927 in the USA under the Hudson River and joined Kennel Street in Manhattan with the 12th and 13th streets of the west suburb of New York, Jersey City. This event uncovered a new direction of underground transport infrastructure development and became an integral part of modern urbanized countries.

One of the modern examples that should be noted is the large-scale project of transferring the main car passageways in Beijing to car tunnels and thus disposing of a considerable part of ecologically harmful car exhaust. The underground passageways will probably approach the surface ones in the number of traffic lines, which in China reach 50. An important aspect of these projects is their consideration for future metropolis development and outpacing the predicted problems of overpopulation.

Another impressive project is Elon Musk's project regarding automobile express tunnels, the first of which was constructed under the Horton City in Los Angeles suburb (2 km long). The idea is an express (240 km/h) transport flow of cars on special platforms that is capable of solving transport and ecological metropolis problems, as the car exhaust volumes decrease significantly. Two ramified networks of tunnels are planned in the Los Angeles region and in the East Coast capital region in the USA. The designer is Elon Musk's The Boring Company.

The most widespread way of mastering downtown underground space is the construction of large trade and entertainment complexes (shopping malls) under the central city squares and adjacent streets. Among the largest we could note: the underground shopping complex "PATH" in Toronto (1,2 thousand stores, 371,6 thousand m^2 underground trading area, a 27 km long network of tunnels, connection to 6 metro stations, over 100 thousand daily visitors); the underground complex of the Town Hall railway station in Sydney (several large underground shopping malls in the 3 km radius of the station, linked by a network of tunnels and passageways); an underground complex in the Albany district in New York (connects three metro stations, bus station, banks, office buildings, shops and restaurants); underground pedestrian center of the Oklahoma City (an enormous pedestrian underground network that connects over 30 shopping malls, museums, theaters, banks, downtown office centers, several metro stations); an underground complex "Stachus" under the Karlsplatz square in Munich (having 5 underground levels and connecting shopping malls with total area over 9 thousand m^2, a railway station and metro stations, car service station, gas station, storage premises and engineering communications, over 100 enterprises; facilitated by 38 escalators). The underground shopping mall construction experience was also successfully applied in the Kyiv city center—the complexes "Globe" (under the Independence Square) and "Metrograd" (Besarabska square and adjacent streets).

Analyzing the future trends of mastering metropolises' underground space, let's study several urban underground construction projects that characterize the urban underground development direction in near and average outlook (Kartosiya 2015; Gilbert, et al. 2013). In Chicago, the USA's second largest economical center, an underground construction of a city is planned, with more than 100 underground floors. Architecturally it will be a 400 m deep "inverted skyscraper" that already got a specific label "earthscraper". To construct this underground complex, 230 million m^3 of ground and rock need to be excavated, and innovative methods of ventilation, lighting, vertical transport, and strong safeguard measures need to be developed. Preliminary assessments put the object's total cost at 15–20 billion USD.

A similar 400 m deep underground complex is planned for construction in Dubai. A special attention here is paid to comfort and safety. A grand hotel at the 350 m depth level is planned. Huge organic material screens built in walls will provide natural lighting, and upon request will broadcast an image of the surface. Each room will have a reserve stock of oxygen, and duplicate exits using adjoined tunnels with moving tracks could quickly transport people to open surface if required. The designed evacuation system relies on 105 elevators with different power supply sources, and their speed (55 km/h) allows to ascend from 400 m depth to the surface in a minute.

The BNKR Arquitectura company designed a project of an "inverted skyscraper" 240 × 240 m in size and 70 floors deep, which is planned to be located under the main square of the Mexico City—Zócalo. The overall complex area totals 775 thousand m^2. The square's surface will be covered by super strong glass that will allow for light passage into the complex, and for passersby to marvel the underground "earthscraper", and located "under foot" museum exhibits of the highest floor (the first 10 levels up to 60 m depth will be given to the Museum of pre-Colombian America). Under the museum level several floors of shopping malls will be placed, below that—a residential block, and starting 180 m deep—business centers. The "inverted skyscraper's" peak will reach 300 m deep.

A grandiose construction project for an underground city in Amsterdam is at the stage of coordination with the authorities and the public organizations of the city. The project AMFORA (Alternative Multifunctional Space Amsterdam) should resolve space, infrastructural, and ecological issues of the Netherlands' capital. As Amsterdam is known for its high density of houses, and a canal system, the project anticipates preliminary draining of canals (which would be restored after finishing construction), and the start of excavation works from the drained canals. A huge volume of underground work indicates that an underground twin city of Amsterdam is planned.

Toronto University's architects devised a project of an underground city in the Nevada desert (the USA) with a dome-like cover. According to the project, the underground structure will remind a honeycomb. Residential and commercial blocks, parks, and gardens are planned in the city. "Honeycomb" cells will be covered by special membranes needed for condensing water from the atmosphere. The underground city will be connected to the Colorado River valley and will improve the water balance and benefit the region's population.

One of the most prospective directions is constructing underground cities in conditions of harsh climates (territories of Canada, Scandinavia, and Siberia). An interesting project of this type is "ECO-CITY 2020" in Myrnyy city, Sakha Republic, Russian Federation, aimed at permanent residence of up to 100 thousand persons. The region is specific, as it has a very low population density (ca. 3 persons per square kilometer) caused by extreme climate conditions: harsh lengthy (6–7 months) winter with an average temperature in January of –35 °C and minimums of –65 °C; hot and short summer (+28 °C; + 36 °C), and very compact transitional periods. To counteract climate troubles the underground city is designed, which is to be located in a crater (quarry) of the exhausted diamond deposit of a kimberlite diamond pipe "Peace" (530 m deep, ca. 1000 m in diameter). The quarry is planned to be covered with a transparent dome with a large number of solar panels (in Myrnyy city region a lot of sunny days in a year are observed).

Due to the ground heat the climate under the dome will be much milder than outside. The space under the dome will be divided into three layers: the bottom layer for farming agricultural products (so-called "vertical farm"); the middle layer is a forest and park zone for purification and enrichment of air; the top layer for permanent residence of people (residential, administrative, commercial and cultural blocks). Natural ventilation of underground space is planned powered by the difference between warm and cold air pressure. The function of heat-saving constructions will be taken upon the rock walls of the quarry, providing payback largely by the high energy efficiency of the complex. For Sakha Republic residents the underground "Eco-city" could become recreational and rehabilitational center, and a peculiar tourist attraction even on global scale. Interesting to know that the project creators analyzed several science fiction ideas, including O. Kazantsev's "The Dome of Hope" (Kazantsev 1980).

The common feature of the considered projects planned to be implemented in various countries and various geological and climate environments is the conceptual vision of an "underground city" as a single giant structure divided into several functional layers and providing to its residents almost all of the comfort and habitual living conditions. It should be noted that the trend of the presented residential objects ("earthscrapers", or "inverted skyscrapers") does not exhaust the full spectrum of complex metropolises' underground space mastering ideas. Mostly the "earthscrapers" are designed as distinct huge, although local, objects, sometimes put outside the city. The cross-sections of these structures are usually significantly less than the whole metropolis area. Thus, these projects should only be viewed as important components of the "underground city" which can expand under the full territory of the metropolis. The fundamental innovation of "earthscrapers" lies in placing the residential blocks underground and in attempts to solve the complicated scientific problem of permanent (long-term) human dwelling in underground space.

Another progressive direction of mastering the urban underground space is the new functional utilization of existing underground structures, i.e., the principle of "postuseful" older objects (Haiko 2018). The largest in-scale example of this technique is the renovation and reuse project of a vast network of old bomb shelters in Beijing, with a total area of nearly 85 km^2. This project will allow to create the largest "underground city" in the world, utilizing the mostly already existing underground space of civil defense shelters that lost their relevance in their current form. However, the new "underground city" can have dual purpose, continuing to serve its previous function of protecting civil population in case of a war.

1.3 The Concept of Sustainable Development of Large Cities and Urban Underground Development

The accumulation of urban, social, and ecological problems of large cities in the second half of the twentieth century required the new ideology of a city development, which led to the concept of "sustainable development" (fully formed in the 1980s). In the "Our common future" of the World Commission on Environment and Development (1987), sustainable development is defined as the development that meets the needs of the present without compromising the ability of future generations to meet their own needs. It is aimed at improving social climate and quality of life, and also on more efficient exploitation of natural resources, and supporting the integrity of the environment (Guiding Principles for Sustainable Spatial Development of the European Continent 2023). The optimal spatial development of large cities is possible only through claiming underground space, which requires a system approach to urban underground planning.

A progressive city building trend of guaranteeing sustainable development that spread through the USA and EU, is the implementation of a "compact city" model, with an important constituent of urban underground development in metropolises.

An essential aspect of sustainable development is the capacity to react to potential changes in the natural environment and minimize technogenic influence. Sustainable urban development considers the factors of economic efficiency, functionality, safety, durability, and the aesthetic of the city in its whole, and also provides the preservation of the cultural and historical architectural legacy. This trend shifts the scope of many engineering projects. The sustainable development methodology dictates that engineers should avoid the traditional envisioning of local tasks in favor of projects in the framework of much larger natural, technical, and social systems. Operation and control within the project should be foreseen for long periods of time, including possibly beyond the designed object's terms of service. This is especially relevant for underground infrastructure which may be in use for centuries. The influence of underground space on the society may be widespread and incredibly useful, and the refusal to utilize it may lead to negative consequences for the city. The underground space could provide the three-dimensional freedom of movement for people, material, water, and energy resources throughout the whole city area.

Millions of people nowadays rely on underground communications that consistently guarantee convenience and comfort. Thus, the exploitation of new underground infrastructure motivates and upholds the sustainable city development, and becomes its integral component. A well-planned and properly maintained underground infrastructure increases the quality of life, energy efficiency, and ecological safety to a much greater extent than a similar surface system.

The implementation of the sustainable development concept for urban underground development requires a systemic resolution of a complex of important tasks (Gilbert et al. 2013; Haiko 2018):

1. Improving the strategic coordination of underground infrastructure development that requires creating the legislative and administrative support of urban underground development. Creating an according administrative body will provide the coordinated planning of urban underground space, preparation of the required constructive norms and regulations; enable collection, archiving, and access regarding the information necessary for decision-making and design. This sort of coordination will also improve the management of scientific research investments, accelerate issuance of permissive documents, and guarantee the support by the state and municipal administrations. The underground space property issues should be clarified at the state legislation level, and transparent decisions regarding the private and municipal property forms of the urban underground space objects should be made.

   As mastering underground space of metropolises is conducted in the subsoil, the legislative regulation is mostly supported by the Constitution and the active laws on subsoil. In the Anglo-Saxon law system (Australia, Great Britain, Canada, the USA, etc.) two legal models are supported: (1) the subsoil belongs to the owner of the surface site, and he freely exploits it; (2) at land sites with no private owner, the government issues the right to the subsoil to the third parties. In most of the countries of continental Europe and South America, the common model is based on the state's property on the subsoil, and it exploits it like a private individual. Here the property rights are divided into the rights on the surface site (usually limited to 2 or 3 m deep) and the subsoil (3 m and deeper).

Ukraine's Constitution, contrary to the most continental European countries, transfers the subsoil property rights, aside from the state, to the local governments also, thus significantly simplifying the exploitation of subsoil for the local communities and encouraging the urban underground development. Additionally, in the Land Codex of Ukraine (p. 1, par. 79) an attempt to glue the "volume" concept to a land site definition is made: "the property rights to a land site extend to the space above and below the ground surface, necessary for construction of residential and industrial structures and facilities". Similar statements are written in the Civil Codex of Ukraine (p. 3, par. 373). In accordance to the Law of Ukraine "Regarding the planning and construction of territories" (par. 24, 27), the feasibility and parameters of constructing underground objects and engineering communications are given.

Despite the certain contradictions between the stated approach and Ukraine's subsoil codex (which was mostly a regulating document for legal relations in the mineral extraction sphere), and also the "flat" understanding of a land site laid down in the land cadaster, the Ukraine's legislation is a progressive factor of urban underground development, although it requires additions and inter-body harmonization for solving some of the issues. To encourage investments in the urban underground development of large cities, the task of creating a three-dimensional land cadaster and clear legal property right attribution on the underground structures becomes especially urgent.

The president of the Associated Research Centers for Urban Underground Space (ACUUS) R. Sterling believes that the property rights on underground space issue are central for urban underground development.[2]

2. The extended and coordinated communication between the involved parties (investors, city and state administrations, geoconstruction organizations, design, scientific and educational centers), and the foundation of a coordinating organization for urban underground development for design, construction, and management of the life cycle of underground objects. A deeper recognition of interrelations between the structural, functional, economic, technological, and natural factors for planning an "underground city", and its interaction with the surface construction and infrastructure is needed. It requires creating the conceptual models of complex mutual impacts between the urban systems (natural and engineering environment, individuals and society, construction machinery and technology, transport, etc.) for understanding the interaction between systems, mitigating the risks, and efficient management in conditions of rapid technology development, societal shifts, and expectations. This analysis should consider positive and negative scenarios, optimizing the interface according to the terms of planning decisions.

[2] R. Sterling's report from the joint UN and International Tunneling Association (ITA) seminar on the sustainable development problems, 14.12.2007.

Attention must be paid to the fact that the decisions regarding the underground infrastructure objects sometimes are made by groups with competing economic or social interests. Devising strategic master plans of urban underground development may soften the acute confrontations, as a planned creation of a large "underground city" will provide enough attractive pieces for a large number of investors, and the approved strategic plan becomes the guarantee of a regional development for investors.

Providing sustainable city development requires interdisciplinary efforts during the whole life cycle of the urban infrastructure. Today the city building science is concentrated in the construction departments; the underground construction—in the mining and transport departments; within the academic science system the urban subject is on the periphery of several disciplines including geography, economics, sociology, and philosophy, complicating the attempts to form the universal theory of urban development. Efficient combination of different knowledge spheres and approaches is seen in the capacities of the modern IT methods that support transdisciplinary research, ontological methods, and applied system analysis.

The work efficiency increases significantly when the design engineers for underground complexes understand the complex social and economic factors, and the city builders have realistic expectations regarding the underground space. Assessing risks under extreme phenomena (terrorism acts, military action, natural disasters, technogenic and ecological catastrophes) may favor the extended mastering of underground space, which is less vulnerable to the consequences of said phenomena. The behavior of people in the conditions of a constantly varying urban environment should also be predicted and taken into account so that the residents can perceive the increased safety of the underground constructions in various complex circumstances.

3. Creating a storage bank as complete as possible for the existing underground structures, and their owner (or leaseholder) companies, and developing the common rules and regulations for monitoring and technical service of the underground structure networks would provide their reliability and safety of functioning. As of now, the information regarding the underground objects of large cities is scattered among different departments and organizations, and no general data bank of the existing and planned underground structures is present; there's a lack of complete scheme maps of the cities' underground space and plans of its development (excluding metropolitan). This state of affairs already caused emergency situations when the construction of the underground object collided with a previous, already existing object. Additionally, no clear division and coordination of such activities as city building and subsoil exploitation is accepted, complicating the rational use of georesources. Underground objects are designed and constructed by separate, disconnected organizations that tend mostly to the personal demands of an object's investor. Traditionally one-time "local" objects are constructed that have no relation to the neighboring underground structures. High diversity of the underground structures and their functions, the technologies of their construction, and the competition between the respective

design and construction organizations hinder the generalization and coordination of the urban underground construction. Even the scientific and educational personnel on geoconstruction are dispersed among different profile universities and research centers (mining, construction, transport), causing divergent methodological approaches to design and construction of underground structures (Kartosiya 2008; Samedov and Kravets 2011; Tajdus et al. 2012).

As the individual owners (organizations) deal with underground structure of different purposes (transport, energy supply, social, engineering, defense, etc.), providing the coordination and management of the whole network of underground excavation should be monitored by an independent service that encompasses all possible types of underground structures with no exceptions.

4. Improving the scientific and educational programs to propagate the modern underground construction technologies and urban underground planning, including new material technologies, artificial intelligence and robotized technologies, laser guidance systems, geoinformation systems, computer models and underground complex visualization techniques, system analysis, and risk minimization methods. Scientific and educational programs should absorb the practical underground construction experience, especially the international experience and technologies, and the advances in other fields of knowledge. An important component of the sustainable urban underground development is the preservation of the best scientific and production collectives after finishing an object, i.e., providing a regular planned manner of urban underground construction. The same is true for the uninterrupted funding of the scientific and educational programs, which will become more interdisciplinary. It may be facilitated by the development of scientific consortiums and inter-department research centers for mastering the urban underground space. As experience shows, the traditional educational problems might satisfy the demands for constructing typical underground objects, but require further professional improvement for implementing the sustainable development concept. Therefore, the programs for master and PhD degrees should give an opportunity to study the international experience and technological advances in both typical and extraordinary large-scale projects.
5. Developing the approaches for assessment of potential threats and risks of mastering underground space. The complete estimate of losses and profits should be conducted during the whole life cycle of the underground objects, i.e., the economic assessment should take into account the long-term perspective, and the ecological and social aspects of the city development. The sustainable system development is proportional to the completeness of information regarding the objective benefits, drawbacks, potential hazards, and risks of the underground infrastructure objects, and their interaction in the city's megasystem, requiring an efficient tool kit for predicting and modeling high-complexity systems, applying system approach techniques. Satisfactory models for integrated large city systems taking into account urban underground construction are not yet developed.
6. Proper accounting of the human factor and the human capabilities of adapting to the underground environment conditions. Underground space may be as safe, attractive, functional, and healthy as the surface space, and in some cases even

more reliable. However, the psychological negative attitude to the "underground" by parts of the society is difficult to overcome. The shift in the negative position will take place with increasing comfort, convenience, design, and reliability of the underground objects compared to the surface ones, when a large number of people will objectively admit the advantages and safety of a long-term stay in the underground environment. Upholding the safety of interacting underground structures of different purposes (e.g., metropolitan and the adjacent multifunctional complexes) requires perfecting fire protection measures, ventilation, structure stability, dynamic impact control, etc. Elaboration of efficient safe behavior rules (not only during construction but also during the operation of underground structures) should complement engineering decisions.

7. Treating the underground space as a valuable non-renewable georesource of a metropolis. It may actively favor the sustainable city development and contribute to the community funds if its understanding, planning, organization, and exploitation are based on the system approach in a long-term perspective, combined with the surface urban development and protection of natural environment. Herewith the planning of the "underground city" should rely on zoning of the metropolis territory by favorability to underground construction using the geological monitoring and geoinformation systems of different purposes. The underground space is not an alternative to the surface; however, it has an untapped potential for infrastructural support of the city's activity, contributing to the sustainable development goals. Political and administrative organs, scientific community, and public organizations should process and agree the long-term concept of the urban underground development that is capable of satisfying the residents' expectations regarding the comfortable living conditions in a metropolis. The support of the scientific research, interdisciplinary education, professional groups of geoconstructors, and "underground city" planners will provide socially, economically, and ecologically attractive development of modern cities according to the state and municipal priorities.

Thus, the placed tasks mean that providing the sustainable urban underground development requires the coordination of efforts of many institutions (state, municipal, manufacturing, finance, scientific, educational, and public), and can be achieved by consolidation of the interested parties in the profile professional or administrative structures. Implementing sustainable development principles will allow to reach the fundamentally new step of mastering underground space, greatly increase the volume of underground construction, and create a ramified multifunctional "underground city"—the integral part of the future metropolises.

References

Ackroyd P (2015) Underground London. Translated from English. O. Morozova Publishing, Moscow (In Russian)

Bryk D, Naryshkin V, Pchelkin V (2012) Regarding the global experience of using underground mining excavations and caves as protective structures. Civil Saf Technol 9(2):22–29 (In Russian)
Diakov S, Kolos O, Verstivskyi A et al (2018) Military fortification structures. Hetman Petro Sahaidachnyi National Academy of Ground Troops, Lviv. (In Ukrainian)
Egorov V, Aksyonov F (1996) Aesthetics of special strategic catacombs. Technics to Youth 2:12–14 (In Ukrainian)
Fink J (1997) Die römischen Katakomben. Philipp von Zabern, Mainz
Frumkin A, Shimron A (2006) Tunnel engineering in the iron age: geoarchaeology of the siloam tunnel, Jerusalem. J Archaeol Sci 33(2):227–237
Gilbert P et al (2013) Underground engineering for sustainable urban development. The National Academies Press, Washington
Golubev G (2005) Underground urbanism and the city. MIKHiS, Moscow (In Russian)
Guiding Principles for Sustainable Spatial Development of the European Continent, rm.coe.int/090000168070018c. Accessed 17 Feb 2023. (In Russian)
Haiko H (2018) Mastering underground space in the concept of sustainable development of large cities. Geotechnologies 1:60–64 (In Ukrainian)
Haiko H, Biletsky V (2015) Mining in the history of civilization: a monograph. Publishing house "Kyiv Mohyla Academy", Kyiv (2015). (In Ukrainian)
Haiko H, Biletsky V, Mikos T, Khmura Ya (2009) Mining and underground structures in Ukraine and Poland (Sketches from History). Donetsk Department of Shevchenko Scientific Society. (In Ukrainian)
Hopkins J (2007) The Cloaca maxima and the monumental manipulation of water in Archaic Rome. The Waters of Rome 4:1–15
Kartosiya B (2008) Introduction to mining science "constructive geotechnology" and the problem of "mastering underground space." MGGU Publishing, Moscow (In Russian)
Kartosiya B (2015) Mastering underground space of large cities. new trends. Sci Inf Anal Bull (sci Tech J) 1:615–629 (In Russian)
Kaufman L, Lysikov B (2009) Large underground cavities: design and construction. Nord-Press, Donetsk (In Russian)
Kazantsev A (1980) The dome of hope. Molodaya Gvardiya, Moscow (In Russian)
Kelemen Ya, Vayda Z (1985) City Underground. Stroyizdat, Moscow. (In Russian)
Lehne M (2008) The complete pyramids. Thames & Hudson, New York
Lykhin P (2003) To history of gradual productivity increase of mining works in tunnel construction. Univ News Min J 1:86–93 (In Russian)
Lysikov B, Kapliukhin A (2005) Utilizing underground space. Nord-Press, Donetsk (In Russian)
Olson A (2012) How Eupalinos navigated his way through the mountain—An empirical approach to the geometry of Eupalinos. Anatolia Antiqua, Institut Français d'Études Anatoliennes (XX):25–34
Passek A (1933) Underwater tunnels. Transzheldorizdat, Moscow (In Russian)
Prepotenska M (2014) Homo urbanus: the phenomenon of a metropolis human. Seredniak T.K. Publishing, Dnipro (In Ukrainian)
Rudniak M (2003) Rational use of underground space for civil objects. MGGU Publishing, Moscow (In Russian)
Rudynskyi M (1961) The stone grave: a pavilion of rock paintings. AN URSR Publishing, Kyiv (In Ukrainian)
Samedov A, Kravets V (2011) Constructing urban underground structures. NTUU "KPI", Kyiv. (In Ukrainian)
Shemyakin E (1995) Utilizing underground space as human habitat. Min J 8:35–39 (In Russian)
Shylin A (2005) Mastering underground space (Genesis and evolution). Mining book, Moscow (In Russian)
Spiro K (1989) Caves of god: cappadocia and its churches. Oxford University Press
State Construction Norms V 2.2.5-97. Civil Defense Protective Structures. State Commission for City Building in Ukraine, Kyiv (1998). (In Ukrainian)

Sterling R, Admiraal H, Bobylev N, Parker H, Godard J, Vähäaho I, Shi X, Hanamura T (2012) Sustainability issues for underground spaces in urban areas. Proc ICE Urban Des Plan 165(4):241–254

Tajdus A, Cala M, Tajdus K (2012) Geomechanika w budownictwie podziemnym. Projektowanie i budowa tuneli. AGH, Krakow

Vähäaho I (2014) Underground space planning in Helsinki. J Rock Mech Geotech Eng 6:387–398

Chapter 2
The Concept of System Approach to Mastering Underground Space of Large Cities

Abstract The concept of a system approach to developing the urban underground construction is unfolded. Types of systems specifying the control of this development process are considered. A natural-technical system "geourbanism—geological environment" is suggested, encompassing various structural and functional, engineering, and natural factors, and their interrelations, for an integrated study of the surface and underground urban space. The components and elements, and the hierarchical levels of this system are considered. A technique is proposed for typification of the geological environment during the mastering of underground space, with the corresponding zoning of the urbanized territories; an example for testing construction sites using this technique is presented. The method of reserving tunnel reliability by controlling bracing parameters is described as a way of taking into account the geological environment variance along the tunnel track in the construction process.

2.1 The Spatial Natural-Technical System "Geourbanism—Natural Environment"

The capability to purposefully influence the development of metropolises greatly aids the resolution of a number of problems that emerged due to the cities' intense growth during the last decades. The tools for this influence (i.e., management) are the general plans, the capital construction plans, the complex transport schemes, the social and economic development programs, and the coordinated construction investment activities by private and government companies. However, the severe complexity, variability, and frequently undefined (or even random) nature between urban structural components significantly adjust the existing plans, call for their improvement using the system approach (Haiko 2014; Korchak 2010; Resin and Popkov 2013; Popkov et al. 1983; Merlen 1977). It can become the basic scientific methodology for solving urban problems, providing the minimal technical and economic risks and rational use of georesources, by strategic planning of the mutual surface and underground urban development.

N. Pankratova et al., *Modeling the Underground Infrastructure of Urban Environments*, The Urban Book Series, https://doi.org/10.1007/978-3-031-47522-1_2

The system approach as a base principle of the system analysis methodology is centered on the notion of a system, which specifies the management entity. A system correctly selected for research can increase the efficiency of the control processes and open up new potential for scientific and technical research. Among the great diversity of the "system" concepts defined in scientific sources related to underground development, let's note the generalized concept of a "geosystem" introduced by the academician K. M. Trubetsky, that describes the totality of natural and artificial objects having the properties of a system which is created or exploited with the purpose of underground development (Trubetskoy 1997). Another definition related to underground development is the representation of a natural-technical system as an engineering structure with a portion of adjacent geological environment that impacts this structure, and has fixed bounds (Lerner and Petrenko 1999). The conditions for underground construction may be defined by the concept of a "lithotechnical" system introduced by H. K. Bondaryk and L. A. Young, that, contrary to a natural (lithologic) one, is controlled by a human and belongs to the class of cybernetic systems (Lerner and Petrenko 1999; Bondarik and Yarg 1990). The operation mode of a natural-technical (or "lithotechnical") geosystem covers two stages: (1) the unestablished mode, inherent to the excavation period or the underground construction period; (2) relatively stable mode, inherent to the consolidation period of the geological environment that contains the underground structure, and the fading of ground displacements and natural disturbance. These stages differ by the human's capability to manage the underlying geomechanical processes.

In geoconstructive technologies (design, construction, and operation of underground facilities) the geosystem "ground stratum—underground structure" has become widely spread, which lies at the foundation of the modern geomechanical vision of the interaction between the engineering structure and the adjacent environment, and allows to calculate the parameters of the designed underground structures accordingly to the ground layer properties (Shashenko and Pustovoytenko 2016; Haiko 2006). To take into account the influence of the underground structures' construction and bracing technologies, the noted system was extended by A. V. Korchak by the "technology" component ("ground stratum—technology—underground structure") that pertinently reflects the interaction between the elements in local objects within the mostly homogeneous ground layers (Korchak 2001). Herewith the geomechanics considers the ground layer (precisely the interaction between the geological environment and the underground structure) as an external impact factor for the structure, and the engineering geology conversely views the structure as an external factor of impact on the geological environment (Bondarik 1981); however, both cases have an understanding of the said subsystems. V. L. Beliaev replaces the term "ground layer" in the urban underground construction to the more appropriate term "geological environment", and considers the system "geological environment—underground structure", however, the first component remains the local surrounding of a separate underground structure (Belyayev 2012).

All of the mentioned approaches to selecting the geosystem and its constituents set an aim of guaranteeing the stability and functionality of a local underground structure and do not consider the large-scale interconnected system in the framework of an "underground city". However, the increase in size and number of the lengthy underground objects, and their incorporation into a complex with chamber-like structures, the properties of surrounding ground layers may vary greatly, therefore the "ground layer" concept should acquire a larger scope to represent the interaction between the objects of a ramified urban underground infrastructure with a variable geological environment. Additionally, the interaction between the surface and underground urban objects should be taken into consideration. This creates an essential difference in representing the impact factors for local and large-scale geosystems; furthermore, their description in the framework of the previously described approaches meets visible contradictions. In (Belyayev 2012), V. L. Beliaev acknowledges the higher hierarchy level of a system, the "underground city", however reduces the analysis to the widely represented "underground structure"; Yu. V. Alekseiev introduces the system "external environment" for the surface urban development, but all of its constituents regard only the surface factors.

Thus, the common methodology of solving the local underground construction tasks becomes insufficiently effective for the planned large-scale mastering of the underground city space, and the existing urban surface development approaches can't be directly applied to the underground structures for reasons of the fundamentally different role of the geological environment. An efficient geosystem should account for the structural and functional links between the surface and underground structures, and the communications network, the variability of the geological environment underlying the metropolis, the geoconstructive technology capabilities, and also to cohere with the city development philosophy and the predicted indicators of the city's demands in various infrastructural spheres.

Correspondingly, the "large scale effect" problem of mastering underground space should find its representation in a new basic geosystem. The authors see as the most advisable entity for planning modern metropolis development, the complex natural-technical system "geourbanism—geological environment" that encompasses the full complexity and diversity of the interrelations between the technical and natural factors of the surface and underground metropolis development (Haiko 2014; Haiko and Bulhakov 2014; Haiko et al. 2015).

The "geourbanism" subsystem (Fig. 2.1), in its turn, consists of two subsystems—the surface urban construction and the urban underground space and represents the spatial organization of the city life (both in horizontal and vertical directions), and also the development and functioning of urban systems on different layers. The goal of the selected subsystems is to display the structural and functional links, and the mutual technogenic influence of the surface and underground city structures.

The basis for the systemic urban underground development of metropolises may be the transit-oriented development (TOD) with advanced "intellectual transport systems" (Resin and Popkov 2013; Haiko et al. 2015; Calthorpe 1993). The centers of a transit-oriented project in the underground transport network become the metro and underground railway stations surrounded by the relatively dense underground structures: multifunctional complexes, trade and entertainment centers, storage facilities, garages, parking lots, sports complexes, civil defense objects, etc. The density of underground construction decreases with the distance from these centers. The

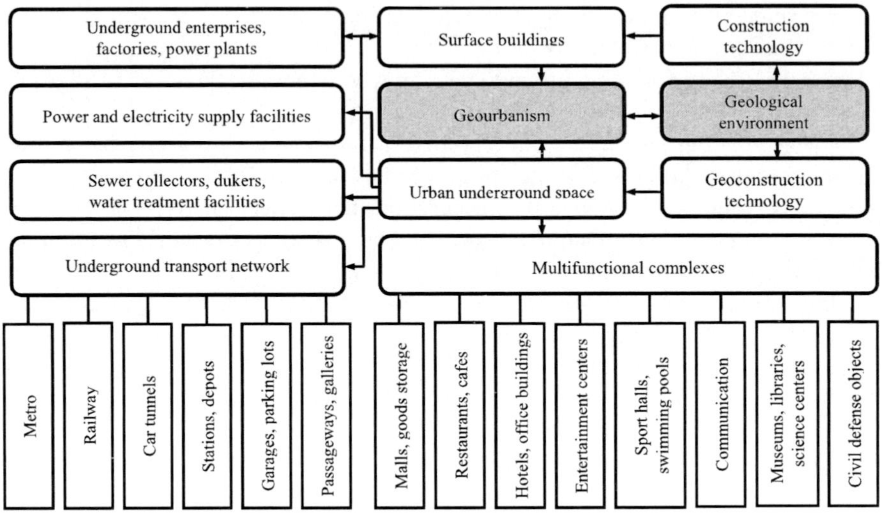

Fig. 2.1 Natural-technical system "geourbanism—geological environment"

urban underground development model for a metropolis includes the "framework" (the main structure-bearing part of a system that eclipses the maximum functional activity concentration area) and the "tissue" (spatial substrate of the system, structurally subordinate to the "framework", that does not require high functional concentration). The framework is formed by the key transport arteries, communication nodes, and connected building complexes that attract visitor flows. Among other things, this model takes into account the permanent metro and car tunnel service of the so-called "pendulum migrations" (daily influx-reflux population movement from the outskirts to the center and back).

Applying the transit-oriented systemic model for planning urban underground space should additionally cover a number of important planning features. The first one is the downtown factor. This metropolis part is the continuous functional activity concentration area, so it requires centralization and the "framework" density accumulation specifically in the downtown area (forming the "core"). This influence factor does not change the nature of the transit flow placement, as it depends on the population density which can be larger in the residential city areas than in the historical downtown area. However, the downtown underground space remains the most active zone for underground construction. For example, in Tokyo, the central city area has up to 16 underground layers (floors) in some sites, where the numerous downtown life activities are located.

On the other hand, the engineering infrastructure is planned according to the population density of a city area (determined by the density and height of surface structures), and the capacity to transfer urban life support systems from surface to underground objects (water and energy supply, sewage, garbage disposal, telecommunication network, etc.). A separate planning task is the underground placement of heavy industrial enterprises located within the city (plants, factories, and power plants).

The second component of the chosen system is the geological environment, which also consists of two parts—natural and technogenic blocks (Fig. 2.2) (Haiko and Kril 2015). According to Construction Norms and Regulations (1992), the geological environment is a multicomponent, discrete, dynamic natural system, diversely and intensely interacting with the structures. It is comprised of a system of geological bodies of different levels, varying compositions, tectonic disturbance, and hydration; they are subdivided into formations, subformations, stratographic lithologic complexes, petrographic types (batches, masses), and mononatural systems.

In the general definition considered in studies (Gorbatiuk 2003; Kril 2002), the geological environment encompasses the spatial section occupied with the geological bodies, bounded at the top by the day surface, and at the bottom by the surface that divides the rock mass altered by any composure parameter, by physical or mechanical, chemical, or other properties as a result of direct or indirect human activity, from the rock mass that did not undergo such changes.

A peculiarity of the metropolis' geological environment is that, in its underground space, the natural geological formations are in part replaced by surface and underground structures, and anthropogenic sediments. The present engineering geological processes are subject to both natural and technogenic factors. The geological environment on the urbanized territories is expanding through deepening of engineering structures and construction of underground complexes that is accompanied by displacements of ground masses, and emergence of new geological conditions.

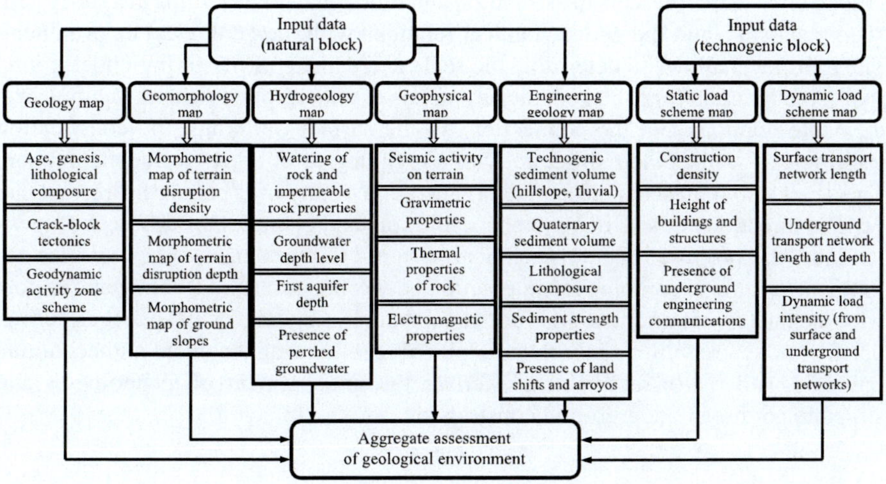

Fig. 2.2 Database structure for assessing geological environment when developing underground space

Active usage of underground space impacts the rock masses (soils) that serves as the foundation. As a result, an area is formed where surface and underground structures, and rock masses (soils) interact; within this area, the state and properties of ground may change, as well as the geohydrological conditions.

The interactions in a natural-technical system "Geourbanism—Geological environment" may occur at three hierarchical levels. The first (lowest) level is a local complex of underground structures and the containing rock (soil) mass. By its nature, this level corresponds to a basic geosystem for local objects with mostly homogeneous engineering geological conditions. The elements of this geosystem are the separate excavations, engineering structures, fastening structures, construction materials. Thus, the first level is a separated volume of underground space that unites the underground excavation (or a complex of it), and the adjacent rock mass into a common functional and geomechanical system of a local type.

The second (middle) level is the city's underground space and its geological environment formed by a combination of separate underground structure complexes into a single "underground city" system. The internal integrity of this type of a system is determined by the presence of stable and persistent structural links among the underground multifunctional complexes, and the geomechanical interaction of engineering objects with the geological environment. This interaction is characterized by the spatial and temporal variability theory (Bondarik 1981), that unlike the local object construction technique considers the variability of construction and operation conditions for the underground structure complexes right from the start, and allows to predict these conditions in the bounds of a metropolis, and to select the most favorable terrains for underground structure placement judging by the functional reliability and safety of structures.

The third (upper) level is the interaction of the surface buildings and the "underground city" that represents the structural and functional relationships in a subsystem "geourbanism", and the geomechanical relations with the geological environment. The third level allows to determine the scale and density of mastering underground space, the functions and the general parameters of underground structure complexes, their interrelations and the connections to the surface buildings, the construction priorities, as well as their structure with a high degree of object conceptualization. This level also represents the spatial organization of city life (both in horizontal and vertical directions), the urban systems evolution, and operation of different scopes.

Thus, the proposed natural-technical system "Geourbanism—Geological environment" is able to combine the interaction between natural and technogenic factors with the factors of structural and functional nature of underground space mastering, opening new possibilities for strategic planning, providing technical and economic rationale, and risk assessment, as well as the improvement of technologies and structures of urban underground construction.

2.2 Determining the Type of Geological Environment and Zoning of Urbanized Territories for Mastering Underground Space

To assess the favorability of the geological environment of the urbanized territories for constructing underground complexes, it is advisable to develop the typing and zoning technique that unlike the existing ones (Gorbatiuk 2003; Kril 2002), would take into account the variability of not only the surface layer on the chosen territory (which is a mandatory requirement for surface construction planning) but also the whole mass (volume) of the geological environment.

The system approach to developing a typing technique requires considering the influence of the whole spectrum of natural and technogenic factors on the underground structure's construction efficiency, stability, longevity, and safety, and its interaction with the surface buildings according to the accepted natural-technical system "Geourbanism—Geological environment".

On this basis a new technique was developed, oriented at mastering underground space (Haiko and Kril 2015). The list of the natural factors includes geology type (age, genesis, lithological composure of sediments, and dislocation forms); terrain morphology (surface heights, terrain disruption, and slope steepness); type and composure of soil and rock; their mechanical characteristics, and properties regarding water; geological displacement of terrain; geodynamic situation; hydrodynamic conditions.

The list of technogenic (geourban) factors included: static load (density and height of surface buildings); external dynamic impact (transit flow intensity, industrial vibration, blasting works, etc.); presence and configuration of underground structures (function type, design, size, form); geoconstructive works method; density and state of hydrotechnical and heat supply structures; presence and nature of technogenic sediment.

The typing principle allows to follow the varying properties of the geological environment on the city's territory. By typing geological environment we understand the selection of its zones characterized by different properties of the undergoing geological and technogenic processes, that have different capacities of reacting to the engineering and practical mastering of the area, manifesting in the specifics of interaction between the natural and artificial elements (structures). The selection is made on the basis of the score from the considered impact factors. This approach is universal and allows to combine the natural and technogenic component values for estimating a geological environment zone regarding the favorability of underground construction.

The main factor indicators were evaluated by a score (rating) of 1–4 points (better conditions mean lesser score) according to the qualitative state of the geological environment: 4 points—the geological environment is unfavorable for underground construction (unsafe impacts of natural or technogenic factors are observed in a large portion of a potential structure's site or a tunnel's track, or the construction itself threatens important surface objects or provokes dangerous geological processes); 3

points—insufficiently favorable geological environment (a part of the underground excavation touches the unsafe site, requiring special construction measures and additional expenses); 2 points—favorable geological environment (most hazardous impacts and risky phenomena are absent, and the construction and operation cost is close to average); 1 point—highly favorable geological environment (great expedience and low cost of excavation, minimal expenses on the structure's support and bracing).

Only the first of the described qualitative states rules out the underground construction capacity and requires bypassing the unfavorable spot (volume) from the sides or from down under, or significant material expenses on improving the construction conditions, with separately calculated technical and economic advisability. The rest of the states acknowledge the potential of mastering underground space, however with different construction cost rates, and different risks regarding the design errors of types I and II (too high or too low construction strength margin). The maximum favorability status (1 point) is introduced for the first time in a study like this, and it allows to prevent design errors of type I, when the favorable conditions are erroneously judged as insufficiently favorable, leading to the unjustified material spending for bracing, and more complex and power-consuming excavation measures than necessary. Typical predictions using two or even three states could not necessarily prevent type I errors.

For the impact factors that could be quantitatively assessed (e.g., the mechanical strength of soil, or construction density), the overall range of possible characteristic values was divided into subranges to match the described favorability states, with an accordingly assigned rating. Using the existing standards, normative documents, rock classifications, etc., is advisable in this process. In cases where the numerical assessment of a factor is impossible, only its presence or absence is declared, and the decision regarding the favorability is determined by an expert method.

The more impact factors are considered, the more justified assessment of the engineering geological mastering potential of the underground space is obtained. The total point ranges that characterize a favorability state considering a number of factors are given in Table 2.1.

Let's consider the issue of the geoinformation support of the geological environment typing and zoning. To conduct typing and visualize the results, a geoinformation

Table 2.1 The geological environment favorability gradation for mastering underground space using the sum of factor points

Geological environment type	I_{sum}*	Data bank mark, Type_GS
Highly favorable	$\leq 1{,}5n$**	1
Favorable	$1{,}5n < \boldsymbol{I_{sum}} \leq 2{,}5n$	2
Insufficiently favorable	$2{,}5n \leq \boldsymbol{I_{sum}} < 3{,}5n$	3
Unfavorable	$\geq 3{,}5n$	4

* $\boldsymbol{I_{sum}}$—total sum of points from all factors

** n—number of considered impact factors

approach was used, which called for creating a data bank in a geoinformation system. It can be formally divided into natural and technogenic blocks (Fig. 2.2).

The data bank is an automated information system of centralized storage and collective access to the database. Data banks may include one or several databases, reference books, dictionaries, query libraries, and applications, and the database management system.

To handle the problems of determining varying geological engineering conditions under technogenic loads, the data bank should include the following information sources:

- the Earth remote sensing (ERS) materials concerning the studied area, their processing, and analysis results;
- technical information (data) about the ERS systems and surface measuring equipment;
- the schemes for profiles, routes, boring hole locations, map networks, surface landmarks, water supply networks, etc.;
- data regarding properties of rock (soil) with existing excavations and information of their status;
- surface and remote photometric and spectral measurements, the results of their statistical processing and analysis;
- topographical, geological, and structural maps of different scales and types;
- scientific-technical, industrial, and patent information about the studied dynamic load sources.

The principles of the data bank organization using the geoinformation approach are the following (Haiko and Kril 2015; Gorbatiuk 2003; Kril 2002):

1. Centralization principle, which is the basis for organization and operation of the data bank. Data integration process involves the unification of various data obtained as a general information volume, facilitating the search and processing of interdependent data.
2. The data transfer to the end users is performed from the single data bank.
3. The server functions as the end user's interface. Additionally, download of the selected necessary data to the users' workplaces is provided.
4. The tools for conversion and analysis of data directly address the single data bank, and the obtained results are stored there.

Considering that the factographic geological information is tightly related to the spatial location of an object, the data bank is founded upon several basic modules, one of them being the geographical information system (GIS) on the base of the MapInfo, Excel software, and the database for scientific research information. In MapInfo an object relation data model is used, called geodatabase (GDB), for storing and providing geographical data (both vector and raster).

The objects stored in the geodatabase are a part of the physical model and carry their descriptions in the logical data model. Therefore, a user working with a GDB simultaneously operates two models: physical and logical, enabling the support of not only the geometrical relation between objects but also their connections on the object level.

This model—the geodatabase—is the core model for storing the whole volume of information; it defines the structure and rules for storing different types of objects: spatial, images, vibration measurement results, geodesic measurements, etc. Using GDB allows for high-speed access and efficient work with the stored within data. Data in the GDB is stored as files with different extensions. They also have linked attribute information; however, the vector data visually describing the terrain on the map, serves as the basis.

In this model, the spatial information is combined with the attributes. The graphical data is stored in special indexed binary files optimized for quick view and access, and the attribute data is stored in tables. The relation between these two types of data is provided by the general identifier field. Also, this model permits to use topological interrelations between vector objects, providing the connectivity between the objects.

The objects in GDB are stored in relation tables. Some tables are the collections of objects, while the others are responsible for the relations between the objects and the rules for integrity checks and the attribute domains. MapInfo controls the integrity of tables, and by aid of access objects to the geographical data provides the object-oriented data model.

The geodatabase structure (object class sets, spatial object classes, topology, and other elements) allows to design geographical databases. The SQL language is used for GDB design in MapInfo, and Visual Basic, Power Builder, and Delphi are used as a visual development environment. All input data should be adapted to the Gauss-Krüger, SK-42, Zone 6 or Latitude/Longitude (WGS 84) coordinate systems.

Practical implementation of the developed geological environment typing system for organization of shallow laying underground objects was tested at the site in the Kyiv city center, in Shevchenkivskyi district near Povitroflotskyi overpass, Fig. 2.3. The site area is 2 × 2 km, and the grid step is 100 m.

In this study, six factors were assessed: soil sediment type (N1), technogenic sediment strength (N2), presence of aquifer and perched groundwater (N3), transport vibration level (T1), building density (T2), and building height (T3).

The thematic layers were created for each factor from the natural and technogenic blocks. To determine the geological environment type (Type_GS) for each of the 400 site cells using the designated range (Table 2.1), the ratings were accessed and accumulated (Fig. 2.4) using the MapInfo SQL queries to the thematic maps.

The table and map analysis show that the selected site mostly has favorable conditions for underground construction (71,25% of the total area), or even highly favorable, in northern and northeastern sections of the site (28,75%), Fig. 2.5.

This is caused mostly by low building density (the average value for the site is 20,7%) and height (average number of floors is 3), admissible transport vibration level (53–73 dB), the absence of technogenic sediment on the large portion of the studied area. On the other hand, almost 90% of the area is composed of soils with properties rated 3 (soft water-saturated sediments of Lybid River bed), and only 10%

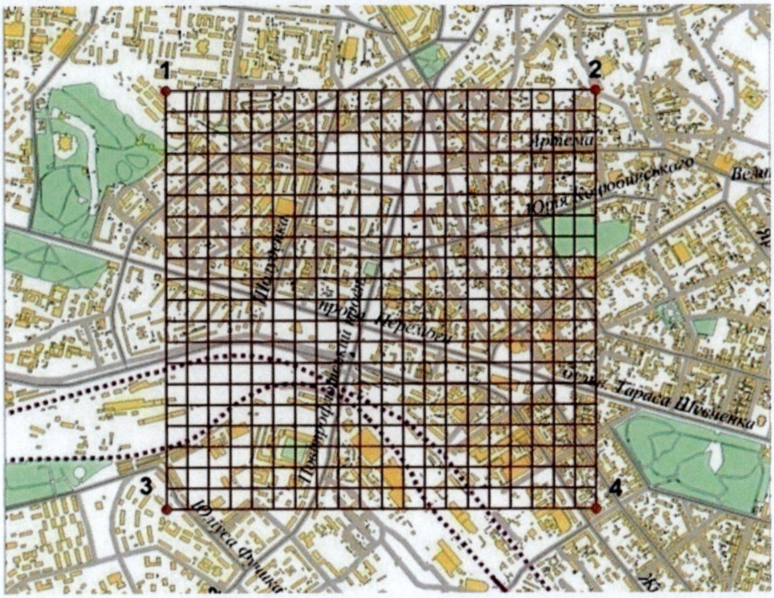

Fig. 2.3 Research area scheme

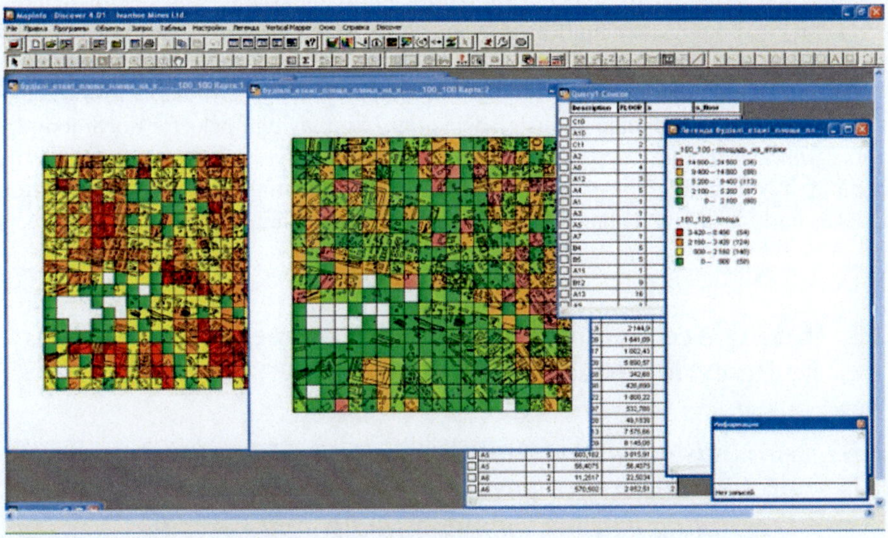

Fig. 2.4 Thematic maps for factors of building density (T2) and height (T3)

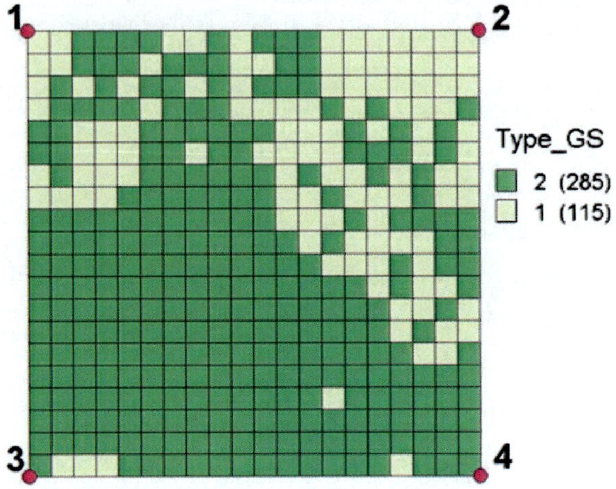

Fig. 2.5 Geological environment favorability assessment scheme for underground space mastering in the selected region of the Kyiv City

is composed of soils favorable for construction—clays and medium-density sands, with alluvial unwatered sediments represented by multi-grained sands.

It should be noted that the technical economic risks from impact of different factors may differ substantially; this should potentially be mitigated by the according weight coefficients, assigned by expert estimation. Particularly the "soil sediment type (N1)" factor should bear a higher influence on the general underground construction favorability assessment.

The proposed geological environment favorability schemes should allow to facilitate and quicken the choice of design decisions (e.g., laying tracks for linear lengthy underground structures bypassing the areas with high *Type_GS* values), assess the complexity and cost of the expected works, and the potential project implementation risks as early as the first stages of technical economic planning.

2.3 Taking Account of Geological Environment Variance by Reserving Tunnel Reliability

2.3.1 Analysis of Variance of Geomechanical Properties of Soils and Active Stress in Tunnel Structures Depending on the Complex Fluidity and Porosity Indicator

Among the important underground influence factors, we should highlight the static load from surrounding soil indicator, which may vary, among other things, when impacted by humidity. The influence of humidity on soils as a geological environment for surface construction objects is studied well enough, and the techniques for its accounting are included in basic design methodology and normative documents (Construction Norms and Regulations 1992; Shvets et al. 2014; State Construction Norms 2014). This is first of all caused by the need to prevent potential soil subsidence under static load from the structure, and also hazardous landslide phenomena. As for underground construction, particularly design of shallow-laying underground structures, the only problem posed was for the soil groups in fluid state, related to the complexity of construction in conditions of drift sand. Other humidity categories weren't viewed as dangerous, lowering the attention to this factor. However, the soil humidity factor can have a significant impact on forming the mountain pressure on the underground construction, as the change in elastic modulus and adhesion of soil leads to an expected increase in displacements and stress (Samedov and Kravets 2011; Tajdus et al. 2012; Haiko et al. 2018). Considering that, let us study the dependence between the strength properties of common soil groups and the changes in humidity.

The influence of humidity is commonly expressed through the fluidity indicator, i.e., the ratio of humidity difference in natural state and on the elastic limit, to the plasticity number:

$$I_l = \frac{W - W_p}{I_p},$$

where W is the natural humidity; W_p is the rolling limit (or, humidity on the rolling limit); and I_p is the plasticity number.

Let us consider the normative distribution of soils by the fluidity indicator I_l for clays and mudrocks (State Construction Norms 2014):

- hard $I_l < 0$;
- semi-hard $0 \leq I_l \leq 0,25$;
- rigid plastic $0,25 < I_l \leq 0,50$;
- soft plastic $0,50 < I_l \leq 0,25$;
- fluid plastic $0,75 < I_l \leq 1$;
- fluid $I_l > 1$.

It should be noted that the strength properties of soils, namely the deformation modulus (E, MPa), and adhesion (cn, kPa), are influenced by the combination of the fluidity indicator (I_l), and the porosity coefficient ($\boldsymbol{e}$), requiring complex consideration of these indicators, and to some extent obstructing the influence analysis for separate dependencies.

To form the database for the geological environment at the right bank Kyiv area, the authors collected and summarized the engineering geological probing materials from the leading construction companies in Kyiv (the localization of respective sites is given in Fig. 2.6).

Data from 362 boreholes (512 geological elements) was reviewed and summarized, with depths ranging from the surface to 60 m deep. The boreholes were situated at the right bank Kyiv area, which is the most prospective for underground development (Haiko et al. 2020).

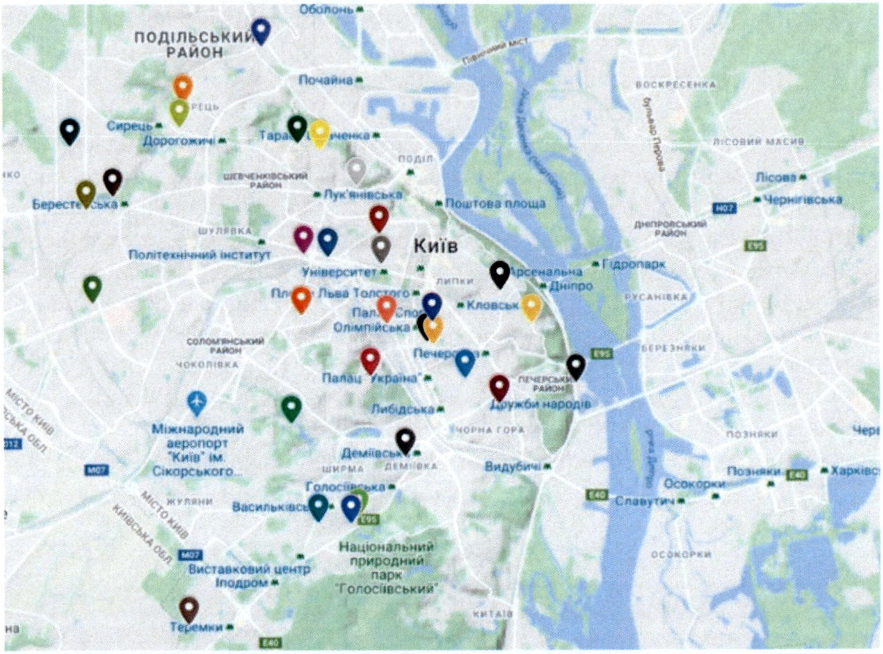

Fig. 2.6 Localization of deep soil probing sites

The fragment of the collected database is given in Table 2.2, where all the geological elements were divided into groups of loamy sands, mudrocks, and clays, and ranged by the fluidity indicator (lowest to highest value). The following characteristics were considered: natural humidity W, rolling limit humidity W_p, plasticity number I_p, fluidity indicator I_l, porosity coefficient $\boldsymbol{e}$, internal friction angle ϕ, specific adhesion cn, and deformation modulus E.

From the obtained data volume, loamy sands, mudrocks, and clays were extracted, for which the surface plots were constructed, setting the deformation modulus (E, MPa) and adhesion (cn, kPa) against the fluidity indicator (I_l) and the porosity coefficient ($\boldsymbol{e}$), respectively; the indicators were grouped according to the accepted ranges (Fig. 2.7).

The obtained dependencies allowed to make a conclusion that all types of soils demonstrate the trend to lower their strength properties (deformation modulus and adhesion) with the increase in humidity. A trend was detected for abrupt (2–4 times) decrease of specific adhesion and deformation modulus for loamy sands and mudrocks, and somewhat milder (1,3–1,7 times) decrease for clays in the range $0 < I_l < 0{,}50$, with further slight decrease in the range $0{,}50 < I_l < 0{,}75$. A similar trend is also observed for the deformation modulus.

Variance in porosity also influences the nature of dependencies but does not change the overall trend of decreasing strength properties of soils. In general, the

Table 2.2 Fragment of the database for geological environment of right bank Kyiv

Object	EGE # (engineering geological element)	Soil name	Natural humidity W	Rolling limit humidity W_p	Plasticity number I_p	Fluidity indicator I_L	Porosity coefficient e	Soil density p, g/cm^3	Internal friction angle ϕ, deg	Specific adhesion cn, kPa	Deformation modulus E, MPa
Peremohy ave., 67	10a	Loamy sands, hard	0,096	0,108	0,04	<0	0,6	1,78	23	14	21
	10b	Sandy loams, plastic	0,161	0,134	0,05	0,54	0,63	1,86	22	11	18
	10c	Sandy loams, fluid	0,189	0,115	0,03	>1	0,65	1,88	18	7	10
	11a	Mudrock, semi-hard	0,172	0,149	0,14	0,16	0,53	2,03	20	34	29
	11b	Light sandy mudrock, rigid plastic	0,177	0,14	0,11	0,34	0,56	1,99	17	25	21
	11c	Light sandy mudrock, soft plastic	0,196	0,134	0,1	0,62	0,59	1,98	16	18	17
	11d	Light sandy mudrock, fluid plastic	0,217	0,132	0,1	0,85	0,61	1,99	13	9	9
	12a	Clays, semi-hard	0,215	0,18	0,22	0,16	0,65	1,95	17	47	24
	21c	Dusty sand, dense	0,183				0,49	2,06	33	5	45

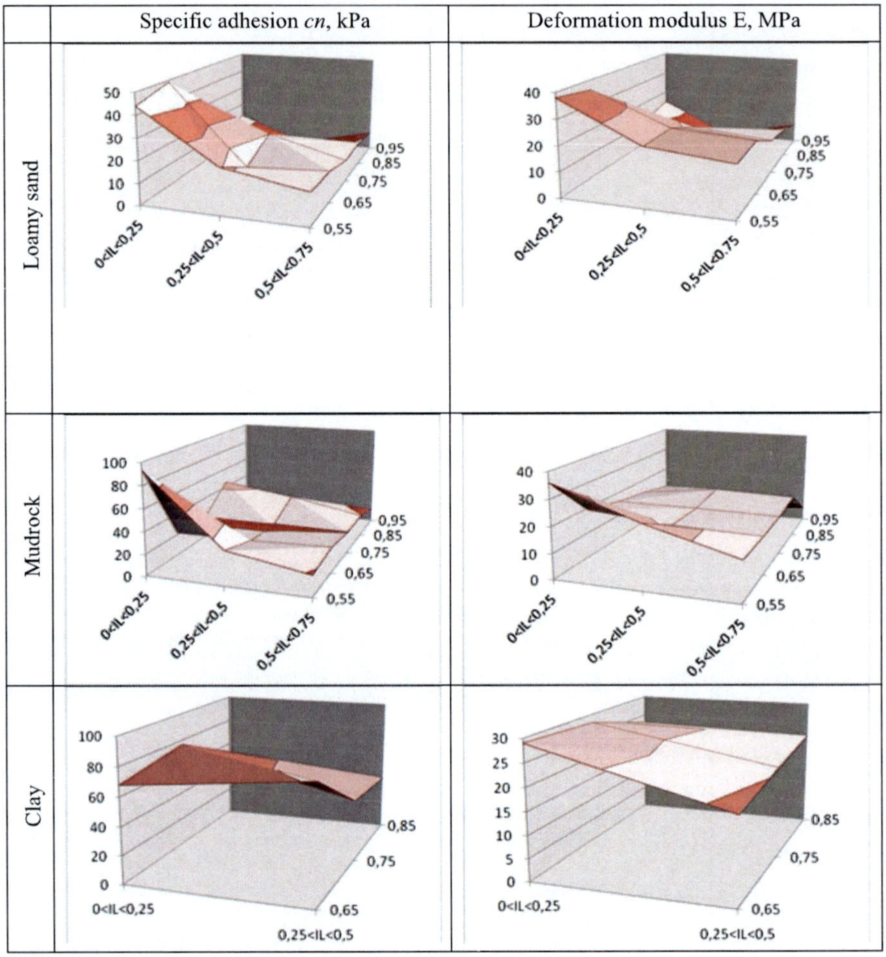

Fig. 2.7 Surface plots of dependencies between the specific adhesion and the soil deformation modulus, and the combination of the fluidity indicator and the porosity coefficient for loamy sand, mudrock, and clay

synthesis of the fluidity indicator I_l and the porosity coefficient ***e*** appeared substantially correct and informative for analysis of variance of mechanical properties in geological environment, with the dependencies in the form of surface plots (unlike two-dimensional graphs) allowing to visibly represent the complexity of the nature of changes, considering the soil structure.

To estimate the influence of humidity change on the displacement of soil around the excavation and the emergence of loads on the casing, a problem was considered for construction of a tunnel in a solid mudrock body with fluidity indicator I_l ranging from 0,1 to 0,75. The tunnel diameter equated to 5 m. The modeled soil body was

30 m wide and 25 m high, with tunnel laying depth 20 m. The problem was modeled as a step-by-step process in PRAXIS (Haiko et al. 2020).

By modeling, the maximum bending moments in the tunnel vault were found in relation to the fluidity coefficient I_l for two tunnel laying depths of 20 and 10 m. The analysis of the obtained data allows to make the following conclusions. All models demonstrate the increase in bending moment in the tunnel vault peak with the growth of fluidity indicator. For loamy sand, the average moment values increase 1,3–1,6 times, for mudrock—1,2–1,4 times, and for clays—1,3–1,5 times. It is worth noting that the distinct combination of factors of porosity and fluidity might form significant deviations from the average values, creating even bigger internal stress in a structure. Comparison of the dependencies for laying depths of 10 and 20 m shows that with the increase in depth the bending moment value increases significantly (1,4–1,8 times, depending on the soil type), however, the impact of humidity preserves the trend of increasing acting moment at the vault peak.

The obtained dependencies show the substantial difference of forming stress and loads in shallow-laying underground structures under varying soil humidity and porosity conditions. This should be taken into account in design decisions by involving probabilistic modeling methods or reserving reliability in a project.

2.3.2 Method of Reserving Tunnel Reliability by Controlling Bracing Parameters

The underground transport network composes a "framework", the core structure-bearing fragment of an "underground city", that encompasses the area of concentration of the most intense functional activity and is formed by the metro tunnel system, as well as the railway, automobile, and engineering tunnels. Under incomplete design information regarding the geological environment properties, an efficient design of a casing (bracing) for a lengthy tunnel becomes one of the most difficult underground construction problems. This is confirmed by the numerous excavation segments in a deformed state, some of which are repaired or refastened with loss of functional operation of a tunnel during the repair time. Simultaneously, there are tunnels where the bracing is underloaded (i.e., has an excessive strength margin), causing the irrational waste of construction material and overpriced construction.

Guaranteeing reliability of lengthy tunnels during their whole service term is a main task in the design and construction of tunnels. Due to the incidental nature of forming rock pressure under conditions of numerous complexly interacting geomechanical factors (like those studied in Sect. 2.3.1 for shallow laying tunnels) that are impossible to account for at the design stage, the tunnel reliability indicator has a probabilistic nature. The most widespread approach in design practice is to determine the estimated load on bracing by the typical adjacent rock displacement indicator (Construction Norms and Regulations, 2004), assuming potential refit and refastening in the segments where the conditions vary from the typical (predicted).

Herewith the expense optimization problem between the increased cost (material spending) of bracing, and the refurnishing works cost may be solved using probabilistic methods from random function theory (Shashenko et al. 2002), or Monte Carlo method (Saługa 2009; Smolich 2004; Haiko and Matviichuk, 2017). However, taking use of probabilistic methods requires a reliable database regarding the excavation state in similar construction and operation circumstances, inhibiting widespread application of the methods.

Beside the traditional approaches to the underground structure design, the two-stage design methods become commonly spread, with their ideas initially proposed by A. Brunner, L. Müller, and L. von Rabcewicz in New Austrian tunneling method (Rabcewicz 1964, 1965). This approach, widely spread and developed by numerous followers for various geological mining conditions, presumes excavation state monitoring, specification, and correction of design decisions during the tunnel construction and fastening, requiring control of the load-bearing capacity of bracing depending on the geomechanical situation along the tunnel, strengthening the structure at the segments where the displacements reach signal values.

The two-stage tunnel design method development with high functional responsiveness lies in the method of excavation bracing with reserve reliability (excess introduction) (Haiko and Gorbatowa 2010; Haiko and Gorbatova 2013a, b), and a new bracing structure with adjusted resistance (Haiko and Gorbatova 2014). The method considers determining the most unfavorable loads on the bracing, and installing through the whole tunnel length an accordingly combined structure (comprised of the permanent and reserve bracing), with its total load-bearing capacity corresponding to the maximum predicted load. At the second stage, an excavation state monitoring is performed, and after the main displacement of ground, in the case of bracing underload, its structure is relieved by gradual dismantling of the reserve bracing elements. The reserve bracing dismantling is performed step by step, with continuous monitoring of rock deformation and combined bracing state. Simultaneously, measures are made for unloading and/or strengthening the adjacent ground mass, requiring the ground-bearing structures as permanent bracing (anchoring combined with spraying concrete, injective strengthening of ground mass, etc.), and support structures as reserve elements (frame bracing, inventory binding elements, etc.), which may be easily removed if not needed.

Let us study the behavior of the system "combined bracing—geological environment" from the reliability theory viewpoint. Let us use the general system equation (Bolotin 1971):

$$Lu = q,$$

where q is an element from the input parameter space Q (external load on bracing from the geological environment);

u is the element from the output parameter space U (bracing characteristic—its load-bearing capacity);

L is the system operator.

Using the bracing characteristics (elements of the U space), any system state may be described, including underloaded, normal operational, unstable, and emergency (an element $u \in U$ corresponds to each of the states). Along the excavation length the states are capable of transforming into each other due to the potential change in the geological environment properties (variance of q values from the input parameter space Q), or the changes in a load-bearing capacity of a bracing (state changes), described by $u(l)$ functions, geometrically displayed as lines in state space U.

To describe the system quality (i.e., the tunnel reliability) we introduce the space V. To each indicator of the underground tunnel reliability (each system quality), an element $v \in V$ corresponds, which is capable of changing along the linear coordinate (tunnel length). Thus, a line $v(l)$ from the quality space V (an underground excavation reliability indicator) corresponds to each line $u(l)$ in the state space U. The relation between the state and quality spaces is as described by the lines of the operator equation:

$$V = Mu,$$

where M is the system operator.

High diversity of the system states that are acceptable from the quality viewpoint (i.e., sufficient tunnel reliability) creates an area of the acceptable states Ω_0 (underloaded and normal operational bracing states) in the quality space V. The bound of the area Ω_0 corresponds to the border state (i.e., the state when the bracing keeps its reliability and remains stable). Let us designate this bound by Γ, and call it the boundary surface (border underground excavation reliability). If $v \in \Omega_0$, then the system quality parameters (excavation reliability indicators) are within the acceptable bounds, and the underground excavation remains stable. Crossing of the boundary surface Γ (acceptable reliability indicator range) by the line $v(l)$ signifies the system failure (tunnel's inability to operate, or excessive risks or costs of operation). The boundary structure state may be presented as an accidental ejection from the acceptable state area.

In the considered system "combined bracing—geological environment" the external influence (external load from the soil on the bracing) $q(l)$ and the equation operator L are stochastic, causing the lines $v(l)$ in the quality space V (underground excavation reliability) to be stochastic too. The system failure (the loss of stability by a tunnel or its segment) is an accidental crossing of the boundary surface Γ by line $v(l)$, i.e., an accidental ejection of the element from the permissible state area. The reliability function in this case will be designated as the probability of the presence of the element $v(\tau)$ in a permissible area Ω_0, within the interval $0 < \tau \leq l$:

$$P(l) = P[v(\tau) \in \Omega_0; 0 < \tau \leq 1].$$

This is the general scheme of the excavation reliability calculation that takes into account the physical, technical, and operational conditions of the system "combined bracing—geological environment" functioning. Let us consider the tunnel stochastic model from this viewpoint more substantially. An ambiguous form of ground contour forming uneven contact loads, variable geomechanical properties of ground along the excavation, the geological environment peculiarities (cracks, water saturation, etc.), determined bracing characteristics during the assembly, and some other factors cause the reliability of the system, and the tunnel's stability, to be different on its segments, and vary along the length of the excavation. As a result of impact of the force fields Q_i (rock pressure manifestation form), the elements of this complex system produce and undergo the loads P_i (Fig. 2.8) of different intensities along the excavation length.

The intensity of the external load from the ground on the bracing $q(l)$ is a random function of the coordinate l (Fig. 2.9a), and the actual calculation of $q(l)$ at the design stage is deemed impossible. The random function $q(l)$ has approximately even expected value and variance throughout the length, i.e., is stationary.

Let us assume that the bearing capacity x must be provided at each ith segment of the tunnel under the most unfavorable conditions of load on the basic and the reserve

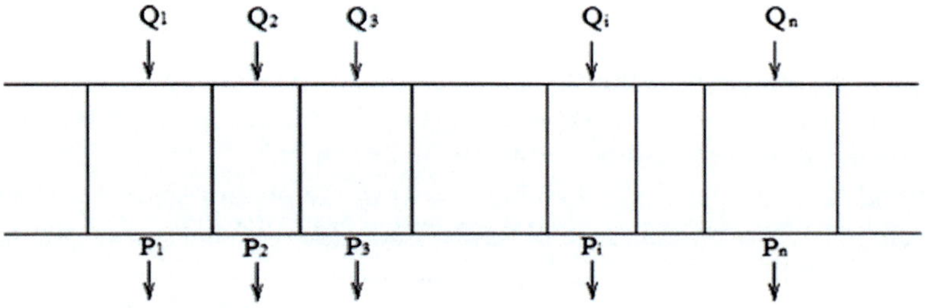

Fig. 2.8 Dependence of the loads P_i that the system "combined bracing—geological environment" produces and undergoes, from the impact of the force fields Q_i

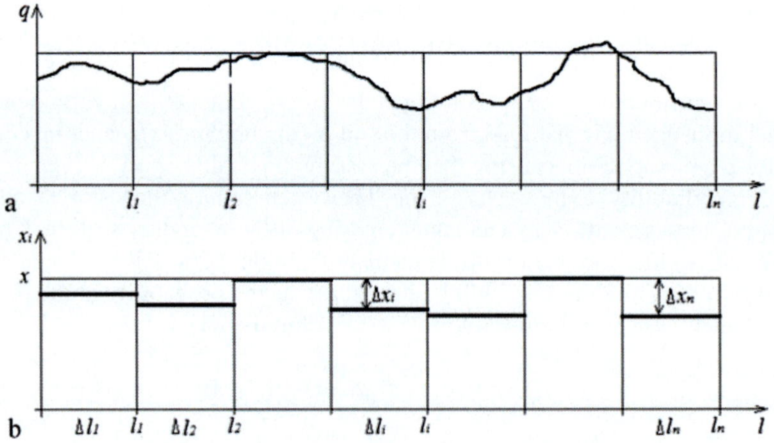

Fig. 2.9 The probabilistic model of an underground excavation fastened by basic and reserve bracing: **a** load distribution q along the excavation length; **b** reserve element values Δx_i that could be dismantled according to the load distribution q

bracing from the geological environment (presence of maximum load). In case an underloaded state of bracing is detected on some segments of the underground excavation during the operation stage, we have Δx_i reserve bracing elements (Fig. 2.9b) that should be designed with a potential for dismantling (to be conformed to the actual load).

The tunnel reliability indicator will be expressed as

$$R(\Delta x_1; \ldots; \Delta x_n) = \prod_{i=1}^{n} R_i(\Delta x_i),$$

where $R(\Delta x_1; ...; \Delta x_n)$ is the excavation reliability indicator when the 1st segment has Δx_1 reserve elements, 2nd segment has Δx_2 reserve elements, and so on, nth segment has Δx_n reserve elements; $R_i(\Delta x_i)$ is the reliability indicator for the ith segment of the excavation, if Δx_i reserve bracing elements are present there;

n is the total number of excavation segments ($i = 1, 2, ..., n$).

We will define the bearing capacity of the bracing that provides the target excavation reliability as

$$W(\Delta x_1; ...; \Delta x_n) = \sum_{i=1}^{n} W_i(\Delta x_i),$$

where $W_i(\Delta x_i)$ is the bearing capacity of the ith segment of bracing, if Δx_i reserve bracing elements are present there.

For the ith tunnel segment, the bearing capacity is equal:

$$W_i(\Delta x_i) = \omega_i \cdot \Delta x_i,$$

where ω_i is the bearing capacity of a single reserve element that is used at the ith tunnel segment.

Then, if only one constraining factor is present, an optimization problem can be formed: by reserving system reliability consisting of n segments, ensure that the reliability indicator is no less than the set value R_0, with the minimum possible bearing capacity (material expense) of the whole underground excavation bracing.

Using the designated notation, the problem can be written as

$$\min\ W(\Delta x_1; ...; \Delta x_n),\ \ R(\Delta x_1; ...; \Delta x_n) \geq R_0.$$

Or, by reserving tunnel bracing reliability that consists of n segments, achieve the maximum possible reliability indicator R, with its material expense (bearing capacity) no more than W_0.

This problem can be written as

$$\max\ R(\Delta x_1; ...; \Delta x_n),\ \ W(\Delta x_1; ...; \Delta x_n) \leq W_0.$$

Tunnel bracing bearing capacity is defined as:

$$W(x_1 + \Delta x_1; \ldots; x_n + \Delta x_n) = \sum_{i=1}^{n} W_i(x_i + \Delta x_i) = \sum_{i=1}^{n} W_i(x_i) + \sum_{i=1}^{n} W_i(\Delta x_i),$$

where $W_i(x_i + \Delta x_i)$ is the bearing capacity of the ith segment of bracing, if x_i basic and Δx_i reserve bracing elements are present there (Fig. 2.9b).

As a bracing reliability indicator on the ith tunnel segment let us introduce the expression:

$$R_{H_i} = \frac{W_i(x_i + \Delta x_i)}{W_0},$$

where W_0 is the maximum load on the most pressured tunnel segment.

Then the reliability indicator throughout the whole length l of the underground excavation can be obtained in the form:

$$R_H = \int_0^l \frac{W_i(x_i + \Delta x_i)}{W_0} dl.$$

Obviously, reliability values closer to 1 express the more rational (efficient) excavation bracing usage.

Contrary to the traditional approach, which applies the strength reserve ratio (critical to actual tension) to determine the bracing efficiency, this new indicator allows to assess the efficiency throughout the whole tunnel length considering the presence of the reserve bracing elements (Δx_i), and the maximum actual loads at the most dangerous segment of the excavation.

From this standpoint let us consider different approaches to bracing design. In traditional design (no reserve elements, $\Delta x_i = 0$) in case of an erroneous lowering of the expected external load (type I design error), a bracing with insufficient bearing capacity will be approved (the proposed R_H indicator significantly less than 1). This leads to bracing failures on numerous segments, and the need to refasten the tunnel, causing continuous operation halt (during the repairs), often inflicting high losses.

In case of an erroneous overestimation of the expected load (type II design error), an underloaded bracing state will be observed, its excess bearing capacity ($R_H > 1$) leads to excessive material spending, redundant expense of work resources, and as a result, an unjustifiably high cost of tunnel bracing.

Now let us consider the design of a tunnel bracing using the controlled (i.e., dismantled when detecting an underloaded state of basic and reserve bracing) reserve elements Δx_i. In this case, a possibility arises to adjust the bearing capacity of the bracing, and advance from $R_H > 1$ to $R_H = 1$ safely for the excavation. As a result, an accordance between the tunnel's bearing capacity and the ground pressure is granted along the whole excavation length, providing minimum material expenses with a guaranteed excavation reliability level.

Thus, the statement of the problem in the form max $R(\Delta x_1; ...; \Delta x_n)$ allows to optimize the economic indicators of a reliable bracing and support of lengthy excavations and provide resource economy in the underground construction. The developed technique can be efficiently applied for long, functionally responsible underground excavations, especially in complex construction conditions and urban tunnel operations. But it also should be noted that the choice of the most favorable site (track) for tunnel construction by a complex of geological environment factors can significantly improve the economic cost of construction and operation of these objects, and minimize expenses on the basic and reserve bracing structures.

References

Belyayev V (2012) Basics of underground city-building. MGSU, Moscow (In Russian)

Bolotin V (1971) Using methods of probability and reliability theories for calculation of structures. Stroyizdat, Moscow (In Russian)

Bondarik G (1981) General theory of engineering (physical) geology. Nedra, Moscow (In Russian)

Bondarik G, Yarg L (1990) Natural technical system and their monitoring. Eng Geol 5:3–9 (In Russian)

Calthorpe P (1993) The next American metropolis: Ecology, community, and the American dream. Princeton Architectural Press

Gorbatiuk N (2003) Zoning of urbanized territories using integral assessment of engineering geological conditions (on the example of Simferopol City): dissertation for candidate of geological sciences degree: 04.00.07. Institute of Geological Sciences, National Academy of Sciences of Ukraine, Kyiv. (In Ukrainian)

Haiko H (2006) Bracing structures for underground construction: reference book. DonDTU, Alchevsk. (In Ukrainian)

Haiko H (2014) Problems of system planning of underground space in large cities. Visnyk NTUU "KPI". Mining Ser 25:35–40. (In Ukrainian)

Haiko H, Matviichuk I, Tarasiuk O (2020) Influence of changes in geological environment on forming loads on shallow-laying underground structures. Geoengineering 2:27–36 (In Ukrainian)

Haiko H, Bulhakov V (2014) Metropolis as a system of surface and underground urban development. In: Quality of mineral resources, pp 315–321. (In Ukrainian)

Haiko H, Gorbatova L (2013a) Estimating reliability of mining excavation in managing bearing load capacity of tunnel support. Development of deposits. Ann Sci Tech Dig 131–136, LLC "LizunovPres", Dnipro (In Russian)

Haiko H, Gorbatova L (2013b) Tunnel support with adjustable resistance for magistral mining excavation with high functional responsibility. Socioeconomic and ecological problems of mining industry, construction and energetics. Sci Disgest 1:203–208, Belarus National Technical University, Minsk. (In Russian)

Haiko H, Gorbatova L (2014) Ukrainian utility model patent #89274, E21D11/14. Bracing with adjustable resistance. Bulletin 7, 2014/04/10. (In Ukrainian)

Haiko H, Gorbatowa, L (2010) Method for drifting the headings with reliable reserved support. Gornictvo i Geoinzynieria, 34(2):283–287, AGH, Krakow

Haiko H, Kril T (2015) Typization of geological environment of urbanized territories in mastering of underground space. In: XIV International scientific practical conference "modern informational technologies for managing ecological safety, natural use, emergency measures", pp 173–180, Kyiv. (In Ukrainian)

Haiko H, Matviichuk I (2017) Statement of the probabilistic problem of assessing impact factors on urban underground structures using Monte-Carlo method. In: Proceeding of the international scientific practical conference "Construction Technologies Development Prospects", pp 57–61, NGU, Dnipro. (In Ukrainian)

Haiko H, Kravets V, Bulhakov V, Haiko Yu (2015) Transport-oriented natural technical geosystem "geourbanism – geological environment". Visnyk NTUU "KPI", Mining Ser 29:18–24. (In Ukrainian)

Haiko H, Matviichuk I, Biletskyi V, Saluga P (2018) Methods of predictive assessment of favorability of geological environment for constructing urban underground objects. V.N. Karazin Kharkiv National University reports, "Geology. Geography. Ecology" Series 48:39–51. (In Ukrainian)

Korchak A (2001) Methodology for design of underground structure construction. Nedra, Moscow (In Russian)

Korchak A (2010) Regarding the system approach to guaranteeing stability of metropolises' underground infrastructure. In: Conference proceedings «Fundamental Problems of Forming Technogenic Geoenvironment», vol 2. Geotechnologies, pp 158–162. IGD SO RAN, Novosibirsk. (In Russian)

Kril T (2002) Influence of technogenic dynamic loads on engineering geological conditions of urbanized territories (on the example of Kyiv City): dissertation for candidate of geological and mineral sciences degree, Kyiv. (In Ukrainian)

Lerner V, Petrenko E (1999) Systematization and perfection of underground objects construction technology. TIMR, Moscow (In Russian)

Merlen P (1977) The city. Quantitative methods of study. Progress. Moscow. (In Russian)

Construction norms and regulations 2.01.15-90 engineering protection of territories, buildings and structures from dangerous geological processes. Basic Principles for Design. Gosstroy, Moscow (1992). (In Russian)

Construction Norms and Regulations II-94–80. Underground Mining Excavation. FGUP TsPP, Moscow (2004). (In Russian)

Popkov Yu, Posokhin M, Gutnov A, Shmulyan B (1983) System analysis and problems of urban development. Nauka, Moscow (In Russian)

Rabcewicz L (1964) The new Austrian tunnelling method. Parts 1, 2. Water Power 11, 453–457, 12, 511–515

Rabcewicz L (1965) The new Austrian tunnelling method. Part 3. Water Power 1:19–24

Resin V, Popkov Yu (2013) Development of large cities under conditions of transitional economy. System approach. Bookhouse «LIBROKOM», Moscow. (In Russian)

Saługa P (2009) Ocena ekonomiczna projektów i analiza ryzyka w górnictwie [Economic Evaluation and Risk Analysis of Mineral Projects]. Studia, Rozprawy, Monografie, nr 152, Wyd. IGSMiE PAN, Kraków

Samedov A, Kravets V (2011) Constructing urban underground structures. NTUU "KPI", Kyiv. (In Ukrainian)

Shashenko A, Pustovoytenko E (2016) Geomechanics: textbook. Novyi druk, Kyiv. (In Russian)

Shashenko A, Tulub S, Sdvizhkova E (2002) Some problems of statistical geomechanics. University publisher "Pulsary", Kyiv. (In Ukrainian)

Shvets V, Boiko I, Vynnykov Yu, Zotsenko M, Petrakov O, Solodiankin O, Shapoval V, Shashenko O, Bida S (2014) Soil mechanics. Groundworks and foundations: textbook. Porohy, Dnipro. (In Ukrainian)

Smolich S (2004) Solving mining geological problems using Monte-Carlo method: reference book. ChitGU, Chita. (In Russian)

State Construction Norms A.2.1-1-2014. Engineering Probing for Construction. Ministry of Regional Construction of Ukraine, Kyiv (2014). (In Ukrainian)

Tajdus A, Cala M, Tajdus K (2012) Geomechanika w budownictwie podziemnym. Projektowanie i budowa tuneli. AGH, Krakow

Trubetskoy K (ed) (1997) Mining science. Mastering and preserving the bowels of the earth. AGN Publishing, Moscow. (In Russian)

Chapter 3
Modified Morphological Analysis Method

Abstract The theoretical foundations of the quantified modification of the classical morphological analysis method—an efficient method for processing uncertain objects with a multitude of possible configurations, are given. The method's place, role, and applications in scenario analysis problems are discussed; types of objects and related uncertainties which the method processes most efficiently, are described. A detailed step-by-step algorithm for the method is given, including main definitions, several procedures involving expert estimation for obtaining input values, the introduction of a numerical cross-consistency matrix, and its significance, several variants of procedures for computing estimates for the alternatives. A two-stage modified morphological method procedure and its multi-stage extensions (networks of morphological tables), aimed at decision-making support for complex systems involving several entities with uncertainties, are presented. A short review of the authors' software used for solving modified morphological analysis problems is given.

3.1 The Place and Role of the Modified Morphological Analysis Method in Scenario Analysis Problems

Managing urban development is one of the most urgent, yet insufficiently researched problems, where rational solutions are particularly hard to find. Spatial distribution of urban objects forms the spatial city structure that defines surface and underground urban development, and the numerous interactions between these objects form the functional city structure. Regulating influences on complex, variable in time and space, often conflicting dependencies in geological, functional, and spatial project configurations requires involvement of special system methodologies.

The development of system analysis as a scientific methodology is tied to the trend of constant complication of systems of different nature—technical, social, economic systems, etc., and as a result, the sophistication of issues and problems that humanity has to deal with in various sphere of its activity. A range of peculiarities

N. Pankratova et al., *Modeling the Underground Infrastructure of Urban Environments*, The Urban Book Series, https://doi.org/10.1007/978-3-031-47522-1_3

in these complex systems makes ineffective or outright impossible the application of traditional operational research methods. These peculiarities include:

- uniqueness;
- no formalized purpose of functioning;
- absence of optimal solution;
- dynamics;
- incomplete description;
- presence of human free will, etc.

The traditional methods of mathematics and statistics mostly extrapolate the retrospective quantitative data to the future, which is satisfactory only for monotonous processes. However, the real-world systems often display qualitative, abrupt changes that have significantly non-linear nature, thus making the quantitative forecast methods inapplicable (Zgurovsky and Pankratova 2005, 2007). That is why the task of envisioning future that is not a simple extension of the present, but acquires fundamentally different forms as a result of certain qualitative changes, requires methodology known as foresight (Zgurovsky and Pankratova 2005). The foresight methodology can be seen as a decision-making process for complex systems with human factor regarding their behavior in the future. This process consists of applying special methods in a certain sequence, with clearly defined links between them (Zgurovsky and Pankratova 2007). It is formed using a broader methodology known as the scenario analysis.

The purpose of scenario analysis is inherently complex due to a number of obstacles, including (Ayres 1969):

- uncertainty, incompleteness, and lack of precision in input data;
- lack of formalization;
- researcher's bias caused by prejudice, inclination to optimistic or pessimistic view of the future, narrow sight of the situation, misrepresentation of facts to fit certain scheme, etc.;
- complexity of interaction between objects in a system—a change in a single object might influence the system as a whole;
- potential capacity of minuscule actions to cause notable consequences (sometimes referred to as the "butterfly effect"). For example, Internet made it possible that an idea or information from a single person may be in mere hours made known to the whole world and cause global impact;
- uncertainty of goals;
- infinite number of possible future variants.

To overcome these obstacles, the scenario analysis methodology is constantly evolving, engaging various qualitative analysis methods, taking into account their advantages and disadvantages, the properties of the researched system in terms of interaction topology between its internal elements, nature of data that circulates in the system (qualitative or quantitative), the conflicts between criteria, and other aspects.

As the noted problems fundamentally can't have a single, optimal solution, the purpose of studying these problems is to find a rational compromise according to the requirements and desires of the decision-maker. To ensure comprehensive, objective, and scientifically grounded analysis of the quality and efficiency of a given strategy in the human/object/environment system, the involvement of system analysts is necessary.

Thus, the definition of system analysis can be given (Zgurovsky and Pankratova 2007): the system analysis is an applied scientific methodology based on a broad diversity of systemically organized, structurally interconnected and functionally interacting heuristic procedures, methodical techniques, mathematical methods, algorithmic and program devices, that provides wholesome, interdisciplinary knowledge about an object of research as an aggregation of interrelated processes of different nature, for further decision-making regarding its evolution and behavior, taking into account the set of conflicting criteria and goals, presence of risks, incompleteness and unreliability of information.

One of the highly productive ways of dealing with complex, unstructured problems is applying the modified morphological analysis method (Pankratova and Savchenko, 2015; Savchenko 2015). The essence of this method is in analyzing objects, processes, or phenomena by their multiple classifications, picking several of their characteristics which can have alternate values. Combining different values for each characteristic, a researcher can produce a very large multitude of object configurations that can be easily analyzed using expert estimations for the constituent parts.

The basic morphological analysis method is quite well known. Its core principles were defined by F. Zwicky (Ritchey and Zwicky 1998; Zwicky 1968; Zwicky and Wilson 1967), and later this method was applied in a number of studies for many different purposes from technical invention (Odrin and Kartavov 1977; Odrin 1986; Akimov 2001, 2005; Levin and Vishnitskiy 2007; Levin and Danieli 2005; Pluzhnikov 1987) to scenario analysis and strategy development (Petrusel and Mocean 2007; Ritchey 2005, 2011, 2006, 2009; Hussain and Ritchey 2011; Isaksson and Ritchey 2003).

The authors developed a special modified morphological analysis method (MMAM) technique specifically for scenario analysis problems (Pankratova and Savchenko 2015; Savchenko 2015, 2011). Its goal is providing decision-making support under circumstances of situational uncertainty, which is one of the defining features of the system analysis problems.

The main application of the developed technique in the scenario analysis deals with the problems that concern objects (processes, events, and phenomena), that by definition have a vast multitude of possible configurations obtained by combining their characteristic features. A source of uncertainty in this case is the inability to determine, which specific configuration of the object in study will appear, so making a rational decision requires study of the whole multitude of possibilities. This uncertainty may be caused by several factors:

1. full information about the object configuration is unknown. This lack of information may be caused by different circumstances, including:

   - the future state of object is studied—after a certain time period or after a specific event happens;

- the decision-maker and analyst have no means of acquiring precise and correct information due to technical, organizational, or other reasons (for example, the problem concerns the surface of other planets or plans of a competing organization).

2. the multitude of possible object configurations are viewed as a whole. In this case, the morphological table describes a vast array of objects with known possible characteristics or constituents, but there is no way to discern which specific combination of them will emerge within the next instance of the object. Examples of objects of this type include traffic accidents in the city's traffic system or profiles of users in some systems;
3. a choice of an object configuration is made. Formally, this case is similar to case 1; however, the uncertainty here is caused not by external factors, and only the decision by the decision-maker resolves this uncertainty.

Modified morphological analysis method can be applied in the scenario analysis process both individually and in conjunction with other methods (Fig. 3.1). At the preliminary analysis stage, the choice of the most rational way of applying MMAM for the problem is made, the characteristic parameters and alternatives are defined. At the qualitative analysis stage, the morphological tables and questionnaires are formed, expert estimation is made, the estimation results are processed, the morphological model is formed and the necessary calculations are conducted; at the scenario writing and selection stage the results of MMAM are involved: the assessments of individual alternatives or object configurations are used for describing the situation in the studied scenario.

3.2 Main Definitions and Guidelines for Constructing Morphological Tables

Let us introduce the main definitions for elements of MMAM.

Definition 1
A characteristic parameter $F_i, \ i \in \overline{1, N}$ of the object of morphological research is a property or attribute that can be used for classification of the variety of objects of the given type.

Generally, a lot of characteristic parameters can be attributed to any given type of objects; the exact selection of characteristic parameters depends on the task and the field of the morphological research.

Definition 2
Alternatives $a_j^{(i)}, \ j \in \overline{1, n_i}$ of a characteristic parameter F_i of the object of morphological research are the mutually exclusive alternative states or values of the respective characteristic parameter.

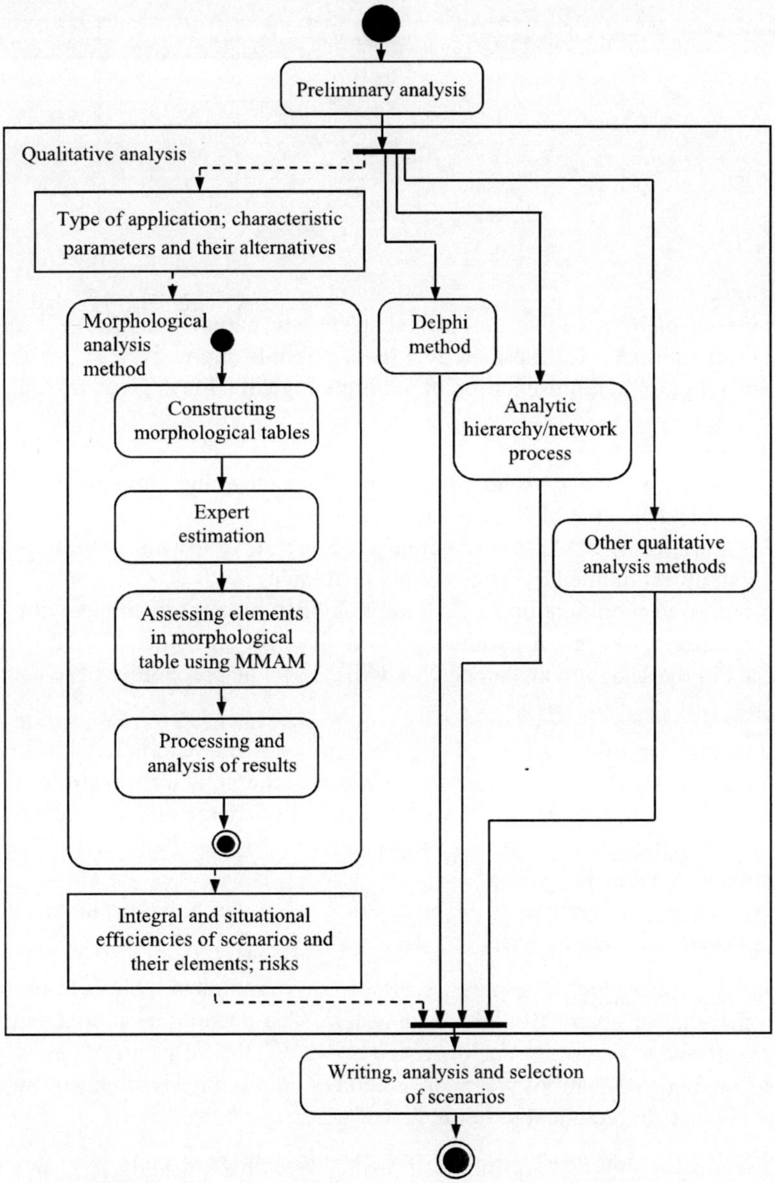

Fig. 3.1 Modified morphological analysis method in the scenario analysis process

Table 3.1 A sample morphological table

F_1	F_2	...	F_N
$a_1^{(1)}$	$a_1^{(2)}$	...	$a_1^{(N)}$
$a_2^{(1)}$	$a_2^{(2)}$	...	$a_2^{(N)}$
...	...	...	...
$a_{n_1}^{(1)}$	$a_{n_2}^{(2)}$	...	$a_{n_N}^{(N)}$

Definition 3
A morphological table (MT) is the set of characteristic parameters $F_i,\ i \in \overline{1, N}$ of an object, each parameter is described by a set of possible alternatives $a_j^{(i)},\ j \in \overline{1, n_i}$.

A general graphic representation of a morphological table is given in Table 3.1.

Definition 4
A configuration s of a morphological table is a set comprising of exactly one alternative for each of the MT's characteristic parameters: $s = \{a_{j_1}^{(1)},\ a_{j_2}^{(2)},\ \ldots,\ a_{j_N}^{(N)}\}$.

A configuration of an MT describes one possible state from a multitude of potential states of an object defined by its morphological table.

The proposed modification of the morphological research operates with probabilities of different states of parameter alternatives for an object of morphological research. For the sake of convenience, we will denote the probability of selecting the alternative $a_j^{(i)}$ as $p_j^{(i)} \equiv P\left(a_j^{(i)}\right)$.

The emergence of one of the alternatives for each characteristic parameter is a guaranteed event that can be represented as a sum of mutually incompatible events—the selection of each given alternative of the respective characteristic parameter. That is why for each parameter F_i the sum of probabilities for emergence of each possible alternative is equal to 1: $\sum_{j=1}^{n_i} p_j^{(i)} = 1$.

Depending on the entities of the morphological research, a morphological table may represent one of three types of a description.

Description of an object. The purpose of such morphological table is to describe a certain material or abstract object, or a system. The uncertainty is inherent to the object itself due to one of the factors noted above. Historically early morphological analysis method applications were concerned with the synthesis of new or improved physical objects or technical systems.

Description of a state. The purpose of such morphological table is to describe a current or future status of some object or, more likely, a system. The object or the system is usually known, so the uncertainty lies in the states of the object's or system's variables that describe what exactly happens with it. A sample case of this research is a description of the state of economy in a chosen future time period. The main parameters of a morphological table that describes a state are the indexes and indicators characterizing the object or system as a whole.

Description of an action (event). The purpose of such morphological table is to describe a specific action or interaction between objects. The system which is the framework for the objects' interaction, is usually known; the uncertainty lies in the exact state of the system (context of the event), the interacting object description, and the characteristics of the event itself. The examples for this type of study include traffic accidents or variants of introducing new product to the market.

A list of typical characteristic parameters includes these:

1. dichotomous ("yes"/"no", "present"/"absent")—the parameters that describe the presence or absence of a certain element or a feature in an object, or an answer to a binary logical question regarding the object. Parameters like these are also necessary when imitating a parameter with multiple-choice alternatives;
2. quantitative (ordinal)—parameters that represent an object's attribute which can be described by a value or an indicator. Sub-ranges of this value comprise the alternatives of such parameter.

This type of parameters can be divided into subtypes by limitations:

- limited—the range for the value is limited on both ends (percentage, probability, etc.);
- unlimited—the range for the value is unlimited at least on one edge (time, profit, quantity, etc.).

This type of parameters can also be divided into subtypes by representation:

- numerical—alternatives are represented as sub-ranges «… to …»;
- verbal—the value of the alternative is described verbally, which is often convenient for non-physical indicators, or in cases when exact ranges are impossible or inappropriate to specify (e.g., alternatives defined as "very small", "small", "average", "large", and "very large");
- comparative—the value is compared verbally to a certain value, which can be a reference, an average, or an expected value (e.g., "less than average", "average", and "more than average").

3. qualitative (nominal)—the alternatives of parameters like these are fundamentally different, and unlike quantitative parameters, the comparative relations between them cannot be set.

For a proper application of MMAM procedure, some rules of MT construction should be adhered to:

- *relevance of parameters*—a characteristic parameter should be interdependent with at least one other parameter (within the level of detail chosen for the problem). This means that the cross-consistency matrix links at least one of the alternatives of this parameter to some other parameter. This rule is not strictly necessary, but ignoring it may create an independent parameter that has no influence on the result, and accordingly, bears no benefit for the research; however, its presence increases load on experts and analysts, and complicates the computational procedure;

- *mutually exclusive alternatives*—as the MMAM algorithms are based on the Bayesian probability apparatus, the states of a single parameter should be mutually inconsistent. If this is not the case, the set of parameters or their alternatives should be redefined to achieve exclusiveness;
- *complete set of alternatives*—each parameter should have a complete set of alternatives so that appearance of one of the parameter's alternatives becomes a guaranteed event. If it is impossible or inconvenient to describe all possible alternatives, an alternative "Other" should be added to the set. If the selection of one of the alternatives is not a necessary event, an alternative "None"/"Not necessary" should be added. More details regarding the completeness of the set of alternatives are given in Savchenko (2022), along with the methods of detecting and handling cases of incomplete alternative sets.

For a specific given task, the morphological table structure is determined by the analysts, often on the base of the results of the preliminary problem study. For analysis of the emerging events in various configurations, the typical groups of characteristic parameters include:

1. contextual parameters—time of event; location of event; circumstances of the event's emergence;
2. reasons and triggers that led to the emergence of the event;
3. specific event characteristics—classification of the event on one or several different axes, often involving the classifications of the involved objects or systems.

The studies regarding the analysis and behavior of a given system have the following groups of characteristic MT parameters:

1. external influence factors with uncertain states at the moment of the system's consideration;
2. characteristics of the studied system's state and behavior.

Rational number of parameters for an MT typically lies in the range of 3–10. In Ritchey (1998), the recommended number of parameters in problems related to foresight is given as not exceeding 6–8; however, by using modern automated calculation procedures, this limit may be raised to 8–10. Larger numbers of parameters often lead to unreasonable numbers of questions to experts, and problems like these often can be decomposed into networks of smaller morphological tables (see Sect. 3.7).

An example of a morphological table for traffic accidents is given in Table 3.2. As can be seen, it contains contextual parameters (2–4), reason parameters (6, 7), and specific event characteristics (1, 5). A configuration of this table bears a single configuration, or scenario, for a traffic accident, and the multitude of configurations (19600 for this table) describes all potential scenarios for a traffic accident, some of which are more or less likely, or outright impossible, and the modification of the classical morphological analysis method is designed to conduct the analysis of this variance.

Table 3.2 Morphological table for traffic accidents

Characteristic parameters						
Type	Place	Time	Weather	Driver's condition	Trigger	Cause
1	2	3	4	5	6	7
Collision with a stationary object	Crossing	Day	Normal	Normal	Excessive speed	Error
Collision of vehicles	Small road	Night	Rain	Alcoholic intoxication	Traffic area violation	Loss of control
Running-down	Wide road		Snow		Red light violation	Conscious violation
Vehicle failure	Tunnel		Icy		Illegal turn	Vehicle malfunction
	Bridge		Fog		Illegal stop	Independent factors
	Parking lot				Illegal area movement	
	Yard				Independent factors	

3.3 Obtaining Input Data in the Formalized Modified Morphological Analysis Method Procedure

Problem statement for expert estimation. The proposed modification of the method sets a purpose of evaluating the probabilities of alternatives for the object described by an MT. The first task is to obtain the initial approximates $p_j'^{(i)}$ for probabilities of alternatives of the characteristic parameters. Ideally, they must be independent probabilities; however, for most real-life problems, this is practically impossible. Therefore, expert estimation is proposed to obtain the assumptions for these values.

Problem statement.

Given:

- a morphological table that contains a set of characteristic parameters $F = \{F_i | i \in \overline{1,\ N}\}$, with each parameter F_i containing a set of alternatives $A_i = \left\{a_j^{(i)} | j \in \overline{1,\ n_i}\right\}$.

Required:

- to obtain the initial approximates $p_j'^{(i)}$ for each of the alternatives $a_j^{(i)}$.

Three methods of obtaining the initial probability approximates are considered:

1. *Equal distribution.* Sometimes adequate assumptions cannot be obtained a priori for various reasons. Significant uncertainty may lead to highly scattered expert opinions regarding the same appraisal. Conversely, expert estimation may be

difficult in the case of proximity of probability values. As the values $p_j'^{(i)}$ are just initial assumptions, it is sometimes rational to skip the initial evaluation completely and distribute an equal probability $p_j'^{(i)} = \frac{1}{n_i}$ for alternatives of each parameter F_i. In this case, the result will be based purely on using cross-consistency matrix for calculation;

2. *Direct expert estimation.* Each alternative $a_j^{(i)}$, $j \in \overline{1, n_i}$ of each parameter F_i, $i \in \overline{1, N}$ is assigned a value $\tilde{p}_j^{(i)}$ by experts using Miller's scale (Pankratova and Malafieieva 2017) (Table 3.3).

The obtained estimates for alternatives are then normalized:

$$p_j'^{(i)} = \frac{\tilde{p}_j^{(i)}}{\sum_{j=1}^{n_i} \tilde{p}_j^{(i)}};$$

3. *Pair-wise expert estimation.* For each parameter F_i, $i \in \overline{1, N}$, the experts assess each pair of its alternatives considering the preference of one alternative over the other in terms of their probability values. Each pair of alternatives $a_j^{(i)}$, $a_k^{(i)}$ is assigned the preference value $m_{jk}^{(i)}$ according to the fundamental scale (Saaty 1993), given in Table 3.4.

If an estimate $m_{jk}^{(i)}$ is given to comparison of jth and kth alternatives, then the symmetrical estimate $m_{kj}^{(i)}$ has inverse value $m_{kj}^{(i)} = \frac{1}{m_{jk}^{(i)}}$.

Table 3.3 Miller's scale for direct assessment of alternatives in a morphological table

Level number	Qualitative level description	Numerical level description
1	Practically impossible	[0 ÷ 0,1]
2	Very low probability	[0,1 ÷ 0,25]
3	Low probability	[0,25 ÷ 0,4]
4	Average probability	[0,4 ÷ 0,6]
5	High probability	[0,6 ÷ 0,75]
6	Very high probability	[0,75 ÷ 0,9]
7	Nearly guaranteed	[0,9 ÷ 1]

Table 3.4 Preference scale

Numerical preference value	Qualitative preference description
1	Equal probability
3	Slight preference
5	Strong preference
7	Very strong preference
9	Absolute preference
2, 4, 6, 8	Intermediate values

Table 3.5 "Accident type" parameter

1. Accident type
1.1. Collision with a stationary object
1.2. Collision of two or more vehicles
1.3. Collision with a pedestrian
1.4. Vehicle malfunction

The probability estimates are calculated using the methods for processing pair-wise comparison matrices—e.g., EM, RGMM, AN methods, etc. (Pankratova and Nedashkivska 2010).

The number of questions to experts is $\sum_{i=1}^{N} n_i$ for direct estimation, and $\sum_{i=1}^{N} \frac{n_i(n_i-1)}{2}$ for pair-wise estimation. Naturally, the pair-wise estimation method not only has the best precision but also increases the number of questions dramatically, especially in studies where characteristic parameters have higher numbers of alternatives. The method of obtaining initial assessments is chosen according to the needs of the study. Generally, the more interdependencies between the parameters are introduced, the less the result depends on initial assessments, so less cumbersome estimation methods may be applied. Direct estimation or even equal distribution is enough for the vast majority of problems.

Let us consider an example of expert estimation for a characteristic parameter "Accident type" from a morphological table that describes traffic accidents (Table 3.5).

Using the equal distribution, we obtain the initial probability estimates $p_j^{\prime(1)} = \frac{1}{n_1} = 0,25$.

In direct expert estimation, the experts are given the questions regarding the estimation for each of the alternatives. A questionnaire may look like this (the expert's answers are underlined):

1.1. Please estimate how frequent are the accidents of the type **«Collision with a stationary object»**

Extremely rare	Very rare	Rare	<u>Common</u>	Frequent	Very frequent	Extremely frequent

1.2. Please estimate how frequent are the accidents of the type **«Collision of two or more vehicles»**

Extremely rare	Very rare	Rare	Common	Frequent	<u>Very frequent</u>	Extremely frequent

1.3. Please estimate how frequent are the accidents of the type **«Collision with a pedestrian»**

Extremely rare	Very rare	Rare	Common	Frequent	Very frequent	Extremely frequent

1.4. Please estimate how frequent are the accidents of the type **«Vehicle malfunction»**

Extremely rare	Very rare	Rare	Common	Frequent	Very frequent	Extremely frequent

The answers are quantified using Miller's scale (Table 3.6), and consequently normalized. Table 3.7 shows the raw and normalized estimates.

Using the pair-wise estimation method for this parameter requires six comparisons using the scale given in Table 3.4. The results of comparison are presented in Table 3.8. The aggregated weight of each alternative is calculated by various methods of pair-wise comparison matrix processing (Pankratova and Nedashkivska 2010).

Cross-consistency matrix. To account for the links between the MT parameters, a numerical cross-consistency matrix (CCM) for the interdependency between alternatives is introduced. Unlike the binary cross-consistency matrix used in studies by other researchers, this approach allows to process the interaction between alternatives in MT configurations with more precision. According to the developed technique (Pankratova and Savchenko 2008), each pair of alternatives $a_{j_1}^{(i_1)}$, $a_{j_2}^{(i_2)}$ of different parameters F_{i_1}, F_{i_2} is assigned a value $c_{i_1 j_1, i_2 j_2} \in [-1; 1]$ according to the Table 3.9.

Table 3.6 Reference of verbal and numerical estimates on the basis of Miller's scale

Level number	Qualitative level description	Numerical level description
1	Extremely rare	0,05
2	Very rare	0,2
3	Rare	0,35
4	Common	0,5
5	Frequent	0,65
6	Very frequent	0,8
7	Extremely frequent	0,95

Table 3.7 Estimates of alternatives for the "Accident type" parameter

Raw estimates		Normalized estimates	
$\tilde{p}_1^{(1)}$	0,5	$p_1'^{(1)}$	0,270
$\tilde{p}_2^{(1)}$	0,8	$p_2'^{(1)}$	0,432
$\tilde{p}_3^{(1)}$	0,35	$p_3'^{(1)}$	0.189
$\tilde{p}_4^{(1)}$	0,2	$p_4'^{(1)}$	0,108

Table 3.8 Pair-wise comparison matrix and the results of its processing

	1.1	1.2	1.3	1.4	Probability estimates by different methods		
					EM	RGMM	AN
1.1	1	1/3	2	4	0,243	0,246	0,214
1.2	3	1	4	5	0,538	0,535	0,57
1.3	1/2	1/4	1	3	0,149	0,15	0,139
1.4	1/4	1/5	1/3	1	0,07	0,069	0,078

Table 3.9 CCM values explanation

Value	Explanation
−1	Alternatives are mutually exclusive; a configuration with this pair of alternatives is impossible
(−1;0)	Alternatives are partially inconsistent; the presence of one of them in a configuration decreases the probability of the second alternative in a pair
0	Alternatives are independent; selecting one of them into a configuration has no impact on selecting the other one
(0;1)	Alternatives are partially linked; the presence of one of them in a configuration increases the probability of the second alternative in a pair
1	Alternatives are fully linked; selecting one of them into a configuration guarantees selecting the second alternative

Estimating this CCM is achieved by presenting the experts with questions regarding the interdependency type of each pair of alternatives of different characteristic parameters. The number of questions may be high, but a significant portion of them may be excluded as obvious or nonsensical when forming the questionnaires. The exact formulation of the questions may vary. A practically convenient form of a question like this is to propose to assess the credibility of a statement that connects the respective alternatives. The experts' answers are then quantified using the scale given in Table 3.10.

As a result of this procedure, a CCM is formed as shown in Table 3.11. It is assumed that the alternatives in a pair equally influence one another, which is why the CCM matrix is symmetrical, and only half of it is shown for simplification purposes.

Table 3.10 Scale for expert estimation of CCM

Expert's answer	Value
Definitely untrue	[−1 ÷ –0,9]
Most likely untrue	[−0,9 ÷ –0,65]
Likely untrue	[−0,65 ÷ –0,35]
More likely untrue than true	[−0,35 ÷ –0,1]
Can be true or untrue evenly	[−0,1 ÷ 0,1]
More likely true than untrue	[0,1 ÷ 0,35]
Likely true	[0,35 ÷ 0,65]
Most likely true	[0,65 ÷ 0,9]
Definitely true	[0,9 ÷ 1]

3.4 Evaluating the Probabilities of Alternatives and Configurations

As the preliminary probability estimates are only approximate, as they do not take into account the interdependencies between the MT parameters, defined by the CCM. To obtain the final probability values for alternatives, a mathematical problem must be solved.

Problem statement:

Given:

- a morphological table that contains a set of characteristic parameters $F = \{F_i | i \in \overline{1,\ N}\}$, with each parameter F_i containing a set of alternatives $A_i = \left\{a_j^{(i)} | j \in \overline{1,\ n_i}\right\}$.
- independent probability values for each of the alternatives $\left\{p_j^{\prime(i)} | i \in \overline{1,\ N};\ j \in \overline{1,\ n_i}\right\}$;
- interdependency values for each pair of alternatives from different parameters $\{c_{i_1 j_1, i_2 j_2} | i_1, i_2 \in \overline{1,\ N};\ i_1 \neq i_2;\ j_1 \in \overline{1,\ n_{i1}};\ j_2 \in \overline{1,\ n_{i2}}\}$.

Required:

- to calculate the probability $p_j^{(i)}$ for each of the alternatives $a_j^{(i)}$, considering the interdependencies between them.

Let us consider this problem for $N = 2$. The selection of any of the alternatives for a parameter F_i is a guaranteed event that can be presented as a sum of n_i mutually exclusive events, constituted by the selection of each of the parameter's alternatives. Therefore, an equation for full probability of the event can be written down for each of the alternatives in the MT. For example, taking the alternative $a_1^{(1)}$ yields the following expression:

Table 3.11 A sample cross-consistency matrix

		F_1				…	F_{N-1}			
		$a_1^{(1)}$	$a_2^{(1)}$	…	$a_{n_1}^{(1)}$	…	$a_1^{(N-1)}$	$a_2^{(N-1)}$	…	$a_{n_{N-1}}^{(N-1)}$
F_2	$a_1^{(2)}$	$c_{11,21}$	$c_{12,21}$	…	$c_{1n_1,21}$					
	$a_2^{(2)}$	$c_{11,22}$	$c_{12,22}$	…	$c_{1n_1,22}$					
	…	…	…	…	…					
	$a_{n_2}^{(2)}$	$c_{11,2n_2}$	$c_{12,2n_2}$	…	$c_{1n_1,2n_2}$					
…	…	…				…				
F_N	$a_1^{(N)}$	$c_{11,N1}$	$c_{12,N1}$	…	$c_{1n_1,N1}$	…	$c_{(N-1)\,1,N1}$	$c_{(N-1)\,2,N1}$	…	$c_{(N-1)\,n_{N-1},N1}$
	$a_2^{(N)}$	$c_{11,N2}$	$c_{12,N2}$	…	$c_{1n_1,N2}$		$c_{(N-1)\,1,N2}$	$c_{(N-1)\,2,N2}$	…	$c_{(N-1)\,n_{N-1},N2}$
	…	…	…	…	…		…	…	…	…
	$a_{n_N}^{(N)}$	c_{11,Nn_N}	c_{12,Nn_N}	…	c_{1n_1,Nn_N}		$c_{(N-1)\,1,Nn_N}$	$c_{(N-1)\,2,Nn_N}$	…	$c_{(N-1)\,n_{N-1},Nn_N}$

$$p_1^{(1)} = \sum_{j=1}^{n_2} P(a_1^{(1)}|a_j^{(2)})p_j^{(2)}, \tag{3.1}$$

or, considering that for an arbitrary event the following equality is true: $P(A|B)P(B) = P(A \cap B)$ (probability multiplication rule), the expression (3.1) can be rewritten as

$$p_1^{(1)} = \sum_{j=1}^{n_2} P\left(\left\{a_1^{(1)}, a_j^{(2)}\right\}\right),$$

meaning that the probability of selecting $a_1^{(1)}$ equals the sum of probabilities of all the MT configurations that contain the respective alternative.

Using expressions like (3.1), taken for each alternative of each MT parameter, a system of equations can be constructed for identification of the values $p_j^{(i)}$. However, this requires obtaining the coefficients $P\left(a_{j_1}^{(i_1)}|a_{j_2}^{(i_2)}\right),\ i_1 \neq i_2$ first. These coefficients are determined using the present input data, i.e., the independent probabilities $p_j^{\prime(i)}$, and the CCM values $c_{i_1 j_1, i_2 j_2}$.

There is no single correct method of determining these coefficients. However, as the values $P\left(a_{j_1}^{(i_1)}|a_{j_2}^{(i_2)}\right)$ represent conditional probabilities, a number of constraints for them should be introduced:

1. if $c_{i_1 j_1, i_2 j_2} = -1$, then the configuration is impossible, meaning that $P\left(a_{j_1}^{(i_1)}|a_{j_2}^{(i_2)}\right) = P\left(a_{j_2}^{(i_2)}|a_{j_1}^{(i_1)}\right) = 0$;
2. 2 grows monotonously when increasing $c_{i_1 j_1, i_2 j_2}$;
3. $P\left(a_{j_1}^{(i_1)}|a_{j_2}^{(i_2)}\right)$ grows monotonously when increasing $p_{j_1}^{\prime(i_1)}$;
4. normalization constraint: for each alternative $a_{j_2}^{(i_2)}$, and for each parameter $F_{i_1},\ i_1 \neq i_2$, the following expression should be true:

$$\sum_{j_1=1}^{n_{i_1}} P\left(a_{j_1}^{(i_1)}|a_{j_2}^{(i_2)}\right) = 1;$$

5. solution balance constraint: for each alternative $a_j^{(i)}$ there exists a nontrivial set of values $p_j^{(i)}$, that for any pair of alternatives $a_{j_1}^{(i_1)},\ a_{j_2}^{(i_2)},\ i_1 \neq i_2$ satisfies the expression

$$P\left(a_{j_2}^{(i_2)}|a_{j_1}^{(i_1)}\right)p_{j_1}^{(i_1)} = P\left(a_{j_1}^{(i_1)}|a_{j_2}^{(i_2)}\right)p_{j_2}^{(i_2)} = P\left(\{a_{j_1}^{(i_1)}, a_{j_2}^{(i_2)}\}\right). \tag{3.2}$$

One of the methods of calculating $P\left(a_{j_1}^{(i_1)}|a_{j_2}^{(i_2)}\right)$ that complies with the noted constraints, is using the following expression:

$$P\left(a_{j_1}^{(i_1)}|a_{j_2}^{(i_2)}\right) = \frac{p_{j_1}^{\prime(i_1)}(c_{i_1 j_1, i_2 j_2}+1)}{\sum_{j=1}^{n_{i_1}} p_j^{\prime(i_1)}(c_{i_1 j, i_2 j_2}+1)}. \tag{3.3}$$

Accomplishing the constraints 1–4 is fairly obvious. We will show that the constraint 5 is also true for values $P\left(a_{j_1}^{(i_1)}|a_{j_2}^{(i_2)}\right)$ calculated as (3.3). Let us consider a pair of arbitrary alternatives $a_{j_1}^{(i_1)}$, $a_{k_1}^{(i_1)}$ of the parameter i_1, and a pair of arbitrary alternatives $a_{j_2}^{(i_2)}$, $a_{k_2}^{(i_2)}$ of the parameter i_2. Then we construct a system of Eq. (3.2) for these alternatives:

$$\begin{cases} P\left(a_{j_2}^{(i_2)}|a_{j_1}^{(i_1)}\right)p_{j_1}^{(i_1)} = P\left(a_{j_1}^{(i_1)}|a_{j_2}^{(i_2)}\right)p_{j_2}^{(i_2)}; \\ P\left(a_{k_1}^{(i_1)}|a_{j_2}^{(i_2)}\right)p_{j_2}^{(i_2)} = P\left(a_{j_2}^{(i_2)}|a_{k_1}^{(i_1)}\right)p_{k_1}^{(i_1)}; \\ P\left(a_{k_2}^{(i_2)}|a_{k_1}^{(i_1)}\right)p_{k_1}^{(i_1)} = P\left(a_{k_1}^{(i_1)}|a_{k_2}^{(i_2)}\right)p_{k_2}^{(i_2)}; \\ P\left(a_{j_1}^{(i_1)}|a_{k_2}^{(i_2)}\right)p_{k_2}^{(i_2)} = P\left(a_{k_2}^{(i_2)}|a_{j_1}^{(i_1)}\right)p_{j_1}^{(i_1)}. \end{cases} \tag{3.4}$$

Let us show that these equations are linearly dependent. Substituting $p_{j_2}^{(i_2)}$, $p_{k_1}^{(i_1)}$, $p_{k_2}^{(i_1)}$ in turn in each succeeding equation, we obtain

$$\frac{P\left(a_{j_2}^{(i_2)}|a_{j_1}^{(i_1)}\right)P\left(a_{k_1}^{(i_1)}|a_{j_2}^{(i_2)}\right)P\left(a_{k_2}^{(i_2)}|a_{k_1}^{(i_1)}\right)P\left(a_{j_1}^{(i_1)}|a_{k_2}^{(i_2)}\right)}{P\left(a_{j_1}^{(i_1)}|a_{j_2}^{(i_2)}\right)P\left(a_{j_2}^{(i_2)}|a_{k_1}^{(i_1)}\right)P\left(a_{k_1}^{(i_1)}|a_{k_2}^{(i_2)}\right)P\left(a_{k_2}^{(i_2)}|a_{j_1}^{(i_1)}\right)}p_{j_1}^{(i_1)} = p_{j_1}^{(i_1)}.$$

Rewriting the conditional probability values in the form of (3.3), after a number of reductions, we obtain that the coefficient

$$\frac{P\left(a_{j_2}^{(i_2)}|a_{j_1}^{(i_1)}\right)P\left(a_{k_1}^{(i_1)}|a_{j_2}^{(i_2)}\right)P\left(a_{k_2}^{(i_2)}|a_{k_1}^{(i_1)}\right)P\left(a_{j_1}^{(i_1)}|a_{k_2}^{(i_2)}\right)}{P\left(a_{j_1}^{(i_1)}|a_{j_2}^{(i_2)}\right)P\left(a_{j_2}^{(i_2)}|a_{k_1}^{(i_1)}\right)P\left(a_{k_1}^{(i_1)}|a_{k_2}^{(i_2)}\right)P\left(a_{k_2}^{(i_2)}|a_{j_1}^{(i_1)}\right)} = 1,$$

meaning that the system (3.4) has a nontrivial solution. Accordingly, the full system of equations written as (3.2) for all of the alternatives will also have a nontrivial solution.

Now we can continue to seek the final probabilities $p_j^{(i)}$. Let us write down a system of equations in the form (3.1) for an MT with two parameters, also adding normalizing equations for each of the parameters:

$$\begin{cases} p_1^{(1)} = \sum_{j=1}^{n_2} P\left(a_1^{(1)}|a_j^{(2)}\right)p_j^{(2)};\ \ldots;\ p_{n_1}^{(1)} = \sum_{j=1}^{n_2} P\left(a_{n_1}^{(1)}|a_j^{(2)}\right)p_j^{(2)}; \\ p_1^{(2)} = \sum_{j=1}^{n_1} P\left(a_1^{(2)}|a_j^{(1)}\right)p_j^{(1)};\ \ldots;\ p_{n_2}^{(2)} = \sum_{j=1}^{n_1} P\left(a_{n_2}^{(2)}|a_j^{(1)}\right)p_j^{(1)}; \\ \sum_{j=1}^{n_1} p_j^{(1)} = 1;\ \sum_{j=1}^{n_2} p_j^{(2)} = 1. \end{cases}$$

This system contains $n_1 + n_2 + 2$ equations and $n_1 + n_2$ variables. However, taking into account that the set of alternatives is complete for each parameter, the last equation for each of the parameters can be rewritten as a linear combination of all the previous equations, making the last equation for each parameter redundant.

$$\begin{cases} p_1^{(1)} = \sum_{j=1}^{n_2} P\left(a_1^{(1)}|a_j^{(2)}\right)p_j^{(2)}; \ \ldots; \ p_{n_1-1}^{(1)} = \sum_{j=1}^{n_2} P\left(a_{n_1-1}^{(1)}|a_j^{(2)}\right)p_j^{(2)}; \\ p_1^{(2)} = \sum_{j=1}^{n_1} P\left(a_1^{(2)}|a_j^{(1)}\right)p_j^{(1)}; \ \ldots; \ p_{n_2-1}^{(2)} = \sum_{j=1}^{n_1} P\left(a_{n_2-1}^{(2)}|a_j^{(1)}\right)p_j^{(1)}; \\ \sum_{j=1}^{n_1} p_j^{(1)} = 1; \ \sum_{j=1}^{n_2} p_j^{(2)} = 1. \end{cases} \tag{3.5}$$

The system (3.5) may be solved by any of the methods for solving linear systems of equations; however, some techniques may be applied to obtain the solution directly.

Let us denote

$$P_1 = \begin{pmatrix} P\left(a_1^{(1)}|a_1^{(2)}\right) & \ldots & P\left(a_1^{(1)}|a_{n_2}^{(2)}\right) \\ \ldots & \ldots & \ldots \\ P\left(a_{n_1}^{(1)}|a_1^{(2)}\right) & \ldots & P\left(a_{n_1}^{(1)}|a_{n_2}^{(2)}\right) \end{pmatrix}, \ P_2 = \begin{pmatrix} P\left(a_1^{(2)}|a_1^{(1)}\right) & \ldots & P\left(a_1^{(2)}|a_{n_1}^{(1)}\right) \\ \ldots & \ldots & \ldots \\ P\left(a_{n_2}^{(2)}|a_1^{(1)}\right) & \ldots & P\left(a_{n_2}^{(2)}|a_{n_1}^{(1)}\right) \end{pmatrix},$$

$$\vec{x}_1 = \begin{pmatrix} p_1^{(1)} \\ \ldots \\ p_{n_1}^{(1)} \end{pmatrix}, \ \vec{x}_2 = \begin{pmatrix} p_1^{(2)} \\ \ldots \\ p_{n_2}^{(2)} \end{pmatrix}.$$

The system (3.5) will then be shortened to the following:

$$\begin{cases} \vec{x}_1 = P_1\vec{x}_2; \ \|\vec{x}_1\| = 1; \\ \vec{x}_2 = P_2\vec{x}_1; \ \|\vec{x}_2\| = 1. \end{cases} \tag{3.6}$$

The solution of system (3.6) is $\vec{x}_1, \ \vec{x}_2$—normalized eigenvectors that correspond to the eigenvalue $\lambda = 1$ of matrices P_1P_2 and P_2P_1, respectively. As these matrices are positive, the estimate for eigenvalues and Perron's theorem imply the existence of an eigenvector with positive coordinates. This vector may be found by iterative methods.

Let us now consider the general problem with $N > 2$. The expression (3.1) will now look like

$$\begin{aligned} p_1^{(1)} &= \sum_{j_2=1}^{n_2} \sum_{j_3=1}^{n_3} \cdots \sum_{j_N}^{n_N} P\left(a_1^{(1)}\middle|\left\{a_{j_2}^{(2)}, \ a_{j_3}^{(3)}, \ \ldots, \ a_{j_N}^{(N)}\right\}\right)P\left(\left\{a_{j_2}^{(2)}, \ a_{j_3}^{(3)}, \ \ldots, \ a_{j_N}^{(N)}\right\}\right) = \\ &= \sum_{j_2=1}^{n_2} \sum_{j_3=1}^{n_3} \cdots \sum_{j_N}^{n_N} P\left(\left\{a_1^{(1)}, a_{j_2}^{(2)}, \ a_{j_3}^{(3)}, \ \ldots, \ a_{j_N}^{(N)}\right\}\right) = \\ &= \sum_{j_2=1}^{n_2} \sum_{j_3=1}^{n_3} \cdots \sum_{j_N}^{n_N} P\left(\left\{a_1^{(1)}, a_{j_2}^{(2)}, \ a_{j_3}^{(3)}, \ \ldots, \ a_{j_N}^{(N)}\right\}\middle|a_{j_2}^{(2)}\right)p_{j_2}^{(2)}. \end{aligned}$$

Now a value $P\left(\left\{a_{j_1}^{(1)}, a_{j_2}^{(2)}, \ a_{j_3}^{(3)}, \ \ldots, \ a_{j_N}^{(N)}\right\}\middle|a_{j_k}^{(k)}\right), k \in \overline{1, N}$ must be determined. Again, we form the constraints for this value:

1. if there exists at least one alternative $a_{j_m}^{(m)}$, $m \in \overline{1, N}$, $m \neq k$ from a configuration $\left\{a_{j_1}^{(1)}, a_{j_2}^{(2)}, a_{j_3}^{(3)}, \ldots, a_{j_N}^{(N)}\right\}$, that $c_{kj_k,mj_m} = 0$, then $P\left(\left\{a_{j_1}^{(1)}, a_{j_2}^{(2)}, a_{j_3}^{(3)}, \ldots, a_{j_N}^{(N)}\right\} \middle| a_{j_k}^{(k)}\right) = 0$;
2. $P\left(\left\{a_{j_1}^{(1)}, a_{j_2}^{(2)}, a_{j_3}^{(3)}, \ldots, a_{j_N}^{(N)}\right\} \middle| a_{j_k}^{(k)}\right)$ grows monotonously when increasing any of the c_{kj_k,mj_m}, $m \in \overline{1, N}$, $m \neq k$;
3. $P\left(\left\{a_{j_1}^{(1)}, a_{j_2}^{(2)}, a_{j_3}^{(3)}, \ldots, a_{j_N}^{(N)}\right\} \middle| a_{j_k}^{(k)}\right)$ grows monotonously when increasing any of the $p_{j_m}^{\prime(m)}$, $m \in \overline{1, N}$, $m \neq k$;
4. normalization constraint: for any alternative $a_{j_k}^{(k)}$ the following expression should be true:

$$\sum_{j_1=1}^{n_1} \sum_{j_2=1}^{n_2} \cdots \sum_{j_{k-1}=1}^{n_{k-1}} \sum_{j_{k+1}=1}^{n_{k+1}} \cdots \sum_{j_N=1}^{n_N} P\left(\left\{a_{j_1}^{(1)}, a_{j_2}^{(2)}, a_{j_3}^{(3)}, \ldots, a_{j_N}^{(N)}\right\} \middle| a_{j_k}^{(k)}\right) = 1;$$

5. solution balance constraint: for each alternative $a_j^{(i)}$, there exists a nontrivial set of values $p_j^{(i)}$, such that the probability of any configuration does not change regardless of the multiplication order of its component probabilities.

One of the methods of calculating $P\left(\left\{a_{j_1}^{(1)}, a_{j_2}^{(2)}, a_{j_3}^{(3)}, \ldots, a_{j_N}^{(N)}\right\} | a_{j_1}^{(1)}\right)$ that complies with the noted constraints, is using the following expression (the example is for $k = 1$):

$$P\left(\left\{a_{j_1}^{(1)}, a_{j_2}^{(2)}, \ldots, a_{j_N}^{(N)}\right\} \middle| a_{j_1}^{(1)}\right) = \frac{P'\left(\left\{a_{j_1}^{(1)}, a_{j_2}^{(2)}, a_{j_3}^{(3)}, \ldots, a_{j_N}^{(N)}\right\} \middle| a_{j_1}^{(1)}\right)}{\sum_{k_2=1}^{n_2} \sum_{k_3=1}^{n_3} \cdots \sum_{k_N=1}^{n_N} P'\left(\left\{a_{j_1}^{(1)}, a_{k_2}^{(2)}, a_{k_3}^{(3)}, \ldots, a_{k_N}^{(N)}\right\} \middle| a_{j_1}^{(1)}\right)}, \tag{3.7}$$

where

$$P'\left(\left\{a_{j_1}^{(1)}, a_{j_2}^{(2)}, a_{j_3}^{(3)}, \ldots, a_{j_N}^{(N)}\right\} \middle| a_{j_1}^{(1)}\right) = \prod_{m=2}^{N} p_{j_m}^{\prime(m)} \cdot \prod_{m=1}^{N-1} \prod_{l=m+1}^{N} \left(c_{mj_m,lj_l} + 1\right).$$

The general representation of the system of equations for obtaining final probabilities can be presented as (3.8)

$$
\begin{cases}
p_1^{(1)} = \sum_{j_2=1}^{n_2} \sum_{j_3=1}^{n_3} \cdots \sum_{j_N=1}^{n_N} P\left(\left\{a_1^{(1)}, a_{j_2}^{(2)},\ a_{j_3}^{(3)},\ \ldots,\ a_{j_N}^{(N)}\right\} \middle| a_{j_2}^{(2)}\right) p_{j_2}^{(2)}; \\
\ldots \\
p_{n_1}^{(1)} = \sum_{j_2=1}^{n_2} \sum_{j_3=1}^{n_3} \cdots \sum_{j_N=1}^{n_N} P\left(\left\{a_{n_1}^{(1)}, a_{j_2}^{(2)},\ a_{j_3}^{(3)},\ \ldots,\ a_{j_N}^{(N)}\right\} \middle| a_{j_2}^{(2)}\right) p_{j_2}^{(2)}; \\
p_1^{(2)} = \sum_{j_1=1}^{n_1} \sum_{j_3=1}^{n_3} \cdots \sum_{j_N=1}^{n_N} P\left(\left\{a_{j_1}^{(1)}, a_1^{(2)},\ a_{j_3}^{(3)},\ \ldots,\ a_{j_N}^{(N)}\right\} \middle| a_{j_3}^{(3)}\right) p_{j_3}^{(3)}; \\
\ldots \\
p_{n_2}^{(2)} = \sum_{j_1=1}^{n_1} \sum_{j_3=1}^{n_3} \cdots \sum_{j_N=1}^{n_N} P\left(\left\{a_{j_1}^{(1)}, a_{n_2}^{(2)},\ a_{j_3}^{(3)},\ \ldots,\ a_{j_N}^{(N)}\right\} \middle| a_{j_3}^{(3)}\right) p_{j_3}^{(3)}; \\
\ldots \\
p_1^{(N)} = \sum_{j_1=1}^{n_1} \sum_{j_2=1}^{n_2} \cdots \sum_{j_{N-1}=1}^{n_{N-1}} P\left(\left\{a_{j_1}^{(1)}, a_{j_2}^{(2)},\ a_{j_3}^{(3)},\ \ldots,\ a_1^{(N-1)}\right\} \middle| a_{j_1}^{(1)}\right) p_{j_1}^{(1)}; \\
\ldots \\
p_{n_N}^{(N)} = \sum_{j_1=1}^{n_1} \sum_{j_2=1}^{n_2} \cdots \sum_{j_{N-1}=1}^{n_{N-1}} P\left(\left\{a_{j_1}^{(1)}, a_{j_2}^{(2)},\ a_{j_3}^{(3)},\ \ldots,\ a_{n_N}^{(N-1)}\right\} \middle| a_{j_1}^{(1)}\right) p_{j_1}^{(1)}; \\
\sum_{j=1}^{n_1} p_j^{(1)} = 1;\ \ldots;\ \sum_{j=1}^{n_N} p_j^{(N)} = 1.
\end{cases}
\tag{3.8}
$$

Let us introduce the matrices $P_1,\ \ldots,\ P_N$:

$$
P_1 = \left[\sum_{j_3=1}^{n_3} \cdots \sum_{j_N=1}^{n_N} P\left(\left\{a_k^{(1)}, a_l^{(2)},\ a_{j_3}^{(3)},\ \ldots,\ a_{j_N}^{(N)}\right\} \middle| a_l^{(2)}\right)\right]_{k\in 1,n_1,\ l\in 1,n_2},
$$
$$
P_2 = \left[\sum_{j_1=1}^{n_1} \sum_{j_4=1}^{n_4} \cdots \sum_{j_N=1}^{n_N} P\left(\left\{a_{j_1}^{(1)}, a_k^{(2)},\ a_l^{(3)},\ \ldots,\ a_{j_N}^{(N)}\right\} \middle| a_l^{(3)}\right)\right]_{k\in 1,n_2,\ l\in 1,n_3},
$$
$$\ldots$$
$$
P_N = \left[\sum_{j_2=1}^{n_2} \cdots \sum_{j_{N-1}=1}^{n_{N-1}} P\left(\left\{a_l^{(1)}, a_{j_2}^{(2)},\ a_{j_3}^{(3)},\ \ldots,\ a_k^{(N)}\right\} \middle| a_l^{(1)}\right)\right]_{k\in 1,n_N,\ l\in 1,n_1},
$$

and the vectors $\vec{x}_1,\ \vec{x}_2,\ \ldots,\ \vec{x}_N$:

$$
\vec{x}_1 = \begin{pmatrix} p_1^{(1)} \\ \ldots \\ p_{n_1}^{(1)} \end{pmatrix}, \ldots, \vec{x}_N = \begin{pmatrix} p_1^{(N)} \\ \ldots \\ p_{n_N}^{(N)} \end{pmatrix}.
$$

Then the system (3.8) can be shortened similarly to (3.6):

$$
\begin{cases}
\vec{x}_1 = P_1\vec{x}_2;\ \|\vec{x}_1\| = 1; \\
\vec{x}_2 = P_2\vec{x}_3;\ \|\vec{x}_2\| = 1; \\
\ldots \\
\vec{x}_{N-1} = P_{N-1}\vec{x}_N;\ \|\vec{x}_{N-1}\| = 1; \\
\vec{x}_N = P_N\vec{x}_1;\ \|\vec{x}_N\| = 1.
\end{cases}
\tag{3.9}
$$

The solution of the system (3.9) contains $\vec{x}_1,\ \vec{x}_2,\ \ldots,\ \vec{x}_N$—normalized eigenvectors that correspond to the eigenvalue $\lambda = 1$, of matrices $P_1P_2\ldots P_N$, $P_2P_3\ldots P_NP_1$, $\ldots$, $P_NP_1\ldots P_{N-1}$, respectively.

To simplify the coefficient matrices let us introduce the following notation. Let C be the multidimensional matrix of order N, containing the products of the CCM values increased by 1, for all the possible pairs for the respective MT configuration:

$$C_{j_1 j_2 \dots j_N} = \prod_{m=1}^{N-1} \prod_{l=m+1}^{N} \left(c_{mj_m,lj_l} + 1\right).$$

We will also introduce the notation $C_{i_1|i_2}$, where $i_1, i_2 \in [1; N]$ are the parameter numbers, for the matrix C multiplied in the respective dimensions by the independent probability vectors for all other parameters $p_i,\ i \neq i_1,\ i \neq i_2$. As each of these multiplications reduces the dimension of the matrix, the resulting matrix $C_{i_1|i_2}$ is an ordinary two-dimensional matrix. Using these notations, the matrices P_i can be written as

$$P_i = diag(\vec{p}_i) \cdot C_{i|i+1} \cdot diag\left(C_{i|i+1}^T \cdot \vec{p}_i\right)^{-1},\ i \in [1;\ N-1];$$
$$P_N = diag(\vec{p}_N) \cdot C_{N|1} \cdot diag\left(C_{N|1}^T \cdot \vec{p}_N\right)^{-1}.$$

It was shown in (Savchenko 2016) that the solution of the system (3.9) is

$$\vec{x}_i = diag(C_{i|k} \cdot \vec{p}_k) \cdot diag(\vec{p}_i) \cdot \vec{1} = diag(C_{i|k} \cdot \vec{p}_k) \cdot \vec{p}_i,$$
$$i \in [1;\ N],\ k \in [1;\ N],\ k \neq i.$$

Considering how $C_{i_1|i_2}$ were constructed, the solution of system (3.9) can be found in the form of (3.10)

$$p_{j_k}^{(i_k)} = \frac{p_{j_k}^{\prime(i_k)} \sum_{j_1=1}^{n_1} \dots \sum_{j_{k-1}=1}^{n_{k-1}} \sum_{j_{k+1}=1}^{n_{k+1}} \dots \sum_{j_N=1}^{n_N} C_{j_1 j_2 \dots j_N} p_{j_1}^{\prime(i_1)} \dots p_{j_{k-1}}^{\prime(i_{k-1})} p_{j_{k+1}}^{\prime(i_{k+1})} \dots p_{j_N}^{\prime(i_N)}}{\sum_{j_1=1}^{n_1} \dots \sum_{j_N=1}^{n_N} C_{j_1 j_2 \dots j_N} p_{j_1}^{\prime(i_1)} \dots p_{j_N}^{\prime(i_N)}}. \tag{3.10}$$

Using the expression (3.10), a probability of any given configuration can be calculated:

$$P\left(\left\{a_{j_1}^{(1)}, a_{j_2}^{(2)},\ \dots,\ a_{j_N}^{(N)}\right\}\right) = \frac{C_{j_1 j_2 \dots j_N} p_{j_1}^{\prime(i_1)} \dots p_{j_N}^{\prime(i_N)}}{\sum_{j_1=1}^{n_1} \dots \sum_{j_N=1}^{n_N} C_{j_1 j_2 \dots j_N} p_{j_1}^{\prime(i_1)} \dots p_{j_N}^{\prime(i_N)}}. \tag{3.11}$$

As a result, by solving (3.9), we obtain the morphological table, that contains the probabilities of selecting alternatives, taking into account the interdependencies between the MT parameters. These values may be analyzed to determine the most critical states of the parameters for the studied object, to rank these states by probability of emergence, to select the most likely configurations, as well as to use these results as input data for the other methods, or the second stage of the two-stage MMAM procedure.

The ordinary MMAM calculation considers all possible states of the MT parameters, i.e., the undefined alternatives, denoted by $a_0^{(i)}$. However, sometimes it may be necessary to study the object characteristics with specific values of some of its parameters, creating a "what-if" inference model. The MMAM's flexibility allows to fix any parameter or parameters, and obtain the probability distributions for the other parameters. For example, given the morphological study from Table 3.2, sample questions that can be answered using this approach, are:

- which places and reasons are the most likely for traffic accidents related to collisions with pedestrians?
- which types of accidents are the most likely due to conscious violation of traffic rules?
- which reasons are the most likely for collisions of vehicles specifically on a crossroads?
- etc.

Even a small morphological table can produce quite a large volume of useful information within this approach.

An example of a table with fixed parameters is given in Table 3.12.

Let's designate a subset of parameters $F' \subset F$ with fixed states, meaning that the selection of one of their alternatives is guaranteed. The set of indices of fixed parameters will be denoted by B, so $F' = \{F_i | i \in B\}$. Each fixed parameter F_i is linked with the index b_i of its fixed alternative $a_{b_i}^{(i)}$. For this alternative, $p_{b_i}^{(i)} = p_{b_i}'^{(i)} = 1$, and for the other alternatives of this parameter, $p_j^{(i)} = p_j'^{(i)} = 0, \ j \in \overline{1, n_i}, \ j \neq b_i$.

Thus, the morphological table with fixed parameters describes a specific situation $S = \left\{a_{j_1}^{(1)}, a_{j_2}^{(2)}, a_{j_3}^{(3)}, \ldots, a_{j_N}^{(N)}\right\}$; $j_k = 0$, if $k \notin B$; $j_k = b_k$, if $k \in B$. For example, the Table 3.12 corresponds to the situation $S = \left\{a_2^{(1)}, a_1^{(2)}, a_0^{(3)}, \ldots, a_0^{(N)}\right\}$.

The MMAM problem statement with fixed parameters is as follows:

Given:

- the morphological table containing a set of characteristic parameters $F = \{F_i | i \in \overline{1, N}\}$, each parameter F_i described by a set of alternatives $A_i = \left\{a_j^{(i)} | j \in \overline{1, n_i}\right\}$;
- independent probabilities of all alternatives $\left\{p_j'^{(i)} | i \in \overline{1, N}; \ j \in \overline{1, n_i}\right\}$;

Table 3.12 Morphological table with fixed parameters F_1, F_2

F_1	F_2	F_3	…	F_N
$a_1^{(1)}$	$a_1^{(2)}$	$a_1^{(3)}$	…	$a_1^{(N)}$
$a_2^{(1)}$	$a_2^{(2)}$	$a_2^{(3)}$	…	$a_2^{(N)}$
…	…	…	…	…
$a_{n_1}^{(1)}$	$a_{n_2}^{(2)}$	$a_{n_3}^{(3)}$	…	$a_{n_N}^{(N)}$

- cross-consistency values for all pairs of the parameters $\{c_{i_1 j_1, i_2 j_2} | i_1, i_2 \in \overline{1,\ N};\ i_1 \neq i_2;\ j_1 \in \overline{1,\ n_{i1}};\ j_2 \in \overline{1,\ n_{i2}}\}$;

parameter subset $F' \subset F$ with fixed states, and the index set b_i of fixed alternatives for the parameters from the subset F'.

Required:

- to calculate the probabilities $p_j^{(i)}$ for each of the alternatives $a_j^{(i)}$, $i \notin B$.

The general system of Eq. (3.8) is modified for this problem by excluding the equations corresponding to the fixed parameters:

$$\begin{cases} \left\{ p_l^{(k)} = \sum\limits_{j_m,\ m \in \overline{1,N} \cap B,\ m \neq k} \ldots \sum P\left(\left\{a_{j_1}^{(1)}, a_{j_2}^{(2)}, \ldots, a_{j_N}^{(N)}\right\} \middle| a_{j_{k'}}^{(k')}\right) p_{j_{k'}}^{(k')} \right\}, & k \in \overline{1, N} \cap B, \\ & l \in \overline{1, n_k}, \\ \left\{ \sum_{i=1}^{n_k} p_i^{(k)} = 1 \right\},\ k \in \overline{1, N} \cap B, \end{cases} \tag{3.12}$$

where k' is the number of the next non-fixed parameter after k, or the number of the first non-fixed parameter, if k is the last non-fixed parameter; $j_m = b_m$ for $m \in B$; the conditional probabilities $P\left(\left\{a_{j_1}^{(1)}, a_{j_2}^{(2)}, \ldots, a_{j_N}^{(N)}\right\} \middle| a_{j_k}^{(k)}\right)$ are calculated using the technique described in this chapter.

It can be shown that the solution of (3.12) can be obtained using (3.10), (3.11), if the input probabilities are set for fixed parameters as $p_{b_i}^{(i)} = p_{b_i}^{\prime(i)} = 1$ when $j = b_i$, and for the other alternatives of this parameter $p_j^{(i)} = p_j^{\prime(i)} = 0,\ j \in \overline{1, n_i},\ j \neq b_i$.

A simple example shows how significant the influence of fixed parameters may be. Assume we have an MT with initial probabilities (Table 3.13), and the CCM (Table 3.14).

The general solution for this problem is given in Table 3.15.

Let's observe how the results for F_2 and F_3 change if different alternatives of F_1 become fixed (Table 3.16).

Table 3.16 demonstrates the impact of selecting a specific fixed alternative on the behavior of other parameters. Particularly, the most likely alternative of F_3 can change depending on this choice.

As such, using MMAM with fixed parameters provides a powerful analysis tool for objects, processes, and phenomena that allows not only to assess the situation as a whole but also conduct modeling of a situation with specific circumstances, generating multiple variable scenarios by fixing different parameters and alternatives.

Table 3.13 A morphological table with initial probabilities

F_1	F_2	F_3
0,3	0,7	0,2
0,4	0,2	0,3
0,3	0,1	0,5

Table 3.14 The CCM for a morphological table from Table 3.13

		F_1			F_2		
		$a_1^{(1)}$	$a_2^{(1)}$	$a_3^{(1)}$	$a_1^{(2)}$	$a_2^{(2)}$	$a_3^{(2)}$
F_2	$a_1^{(2)}$	0,2	0,5	−0,2			
	$a_2^{(2)}$	−0,8		0,5			
	$a_3^{(2)}$	0,8	−0,5	0,2			
F_3	$a_1^{(3)}$	−0,8	0,5	0,8			0,5
	$a_2^{(3)}$	−0,5	0,5			0,5	
	$a_3^{(3)}$	0,5		0,2	−0,5		

Table 3.15 The solution for the problem without fixed parameters

F_1	F_2	F_3
0,186	0,65	0,256
0,508	0,231	0,361
0,306	0,119	0,382

Table 3.16 Dependency of non-fixed parameters on the choice of the fixed alternative for F_1

F_1	F_2	F_3	F_1	F_2	F_3	F_1	F_2	F_3
1	0,69	0,067	0	0,742	0,281	0	0,474	0,33
0	0,059	0,235	1	0,208	0,445	0	0,373	0,299
0	0,251	0,698	0	0,05	0,247	1	0,153	0,371

3.5 Alternative Approach for Processing Cross-Consistency Matrix

The described above procedure of MMAM over the years was utilized by the author in a number of different morphological studies, including the analysis of delays in the technological process of a large facility; analysis of accidents and traffic jams in the transport system of a big city (Savchenko 2011); study of social disasters (Savchenko 2018); analysis of suitability of sites for underground construction (Pankratova et al. 2018; Haiko et al. 2019); etc.

As these studies incorporated expert estimation, some feedback was gathered from people who participated in the studies. Generally, the questions in MMAM are considered understandable and intuitive after simple instructions, and the results are perceived as appropriate and close to reality by the experts. However, a recent morphological study has revealed a recurrent issue that required tweaking the model parameters. Although adjusting model parameters after the initial assessment is an absolutely normal and common practice, the repetitive nature of adjustments led to believe that there is an underlying issue in some parts of the method itself.

The investigation has shown that the issue originates in the treatment of cross-consistency matrix in MMAM. The flaw, while not severely destructive, can be a bit misleading and confusing for the experts and for the analyst in some certain cases of using MMAM. Considering that there are no single "correct" solutions for real-world problems, and the adequacy of results can be measured only by the experts, the procedures of processing input data should be as realistic as possible. As such, this paper discusses alternate techniques for processing CCM in the method.

As was noted before, there is no single correct method of accounting the CCM values in the calculation. During the development of MMAM the procedure of obtaining result values was evolving and finally transformed into the procedure described in (Savchenko 2016). According to this procedure, CCM is converted into scaling coefficients for configurations of the morphological table that change their probabilities accordingly in a range from –100% to +100%.

It is easy to notice that the +100% probability change does not actually correspond to the description of "1" value, as it does not exclude other combinations with the alternatives from the pair.

This was done consciously for a number of reasons. Theoretically, a value of "1" should yield an infinite scaling coefficient, but such treatment breaks the current MMAM procedure, as a subset of configurations gets infinite probability, while the other configurations (including those that do not even have the alternatives from this pair) get zero probability due to normalization.

Additionally, a "1" value in CCM is a very rare occurrence, as competently designed morphological tables seldom have the need for linked alternatives.

So, after some experimenting, the current procedure of treating CCM values according to scaling coefficients was settled as the most balanced, and morphological studies have shown that this procedure most of the time produces plausible results.

The concern arose when during the testing of morphological model for assessing suitability of urban sites for underground construction (Pankratova et al. 2018; Haiko et al. 2019) some alternatives were getting ratings that were slightly lower than expected by the experts. The discrepancy was very subtle and it was easily fixed by re-evaluating some of the input assessments, but it induced a thorough analysis of the causes that led to it.

It turned out that the issue was observed where the configurations that took a number of counterbalancing values in CCM (e.g., 0,5 and –0,5) were involved. From the expert's point of view, these values lie on equal distance from the center, so they should compensate each other; however, the current treatment of CCM assigns 50% and 150% scaling coefficients which together produce 75% instead of expected 100%. This effect is nearly imperceptible when it happens for a single configuration, but if the morphological table is sufficiently large, and the CCM is diverse, this may influence the result. The worst case happens when the configurations that are susceptible to this flaw include some, but not all, of the parameters (in case all the parameters are affected, the normalization would smooth out the result).

To sum up, this means that negative CCM values bear more weight on the result than identical positive CCM values. In this chapter the methods of evening this misbalance by modifying the CCM evaluation procedure or the calculation algorithm in MMAM.

Adjusting the existing procedure. As the current procedure has shown itself as quite reliable, and produces believable results most of the time, it would be good to retain it as the basic procedure, with only slight modifications.

The logical way is projecting CCM values to $[0, \infty)$ range, and translating the questionnaire results that have equal distance from the center into inverted CCM values. The infinity value would be banned due to its destructive effect on the MMAM procedure. This produces less elegant presentations of CCM, as the opposite sides of likelihood spectrum would significantly differ (e.g., 0,25 and 4). The other approach would be to use a logarithmic scale and place CCM values in $(-\infty, \infty)$ range, but it

Table 3.17 Scenario 1 morphological table

1. Company B's position		2. Company C's position	
1.1. Accept	0,8	2.1. Accept	0,4
1.2. Refuse	0,2	2.2. Refuse	0,6

would be even more confusing for the unprepared participant, as impossible pairs of alternatives are quite common in morphological analysis, which will mean that the matrix would be sprinkled with $-\infty$ figures.

But even with demonstrativeness issues aside, this approach still has the problem with positive values, as the scaling coefficient shows favor for a configuration against the whole field of configurations, so the increase in probability will be more than expected, especially for a large morphological table. This means that this kind of treatment does not fix the balance, it actually swings it the other way, with positive values now bearing more weight than negative. Older experiments showed that such misbalance is more evident, which is why this way of treating CCM values was dropped in favor of the current one.

Proposing an alternate procedure. Newly conducted research has shown that an alternate approach may be used instead of the existing technique of treating CCM values as scaling coefficients.

To understand the proposed procedure, its difference from the current procedure, and the impact on the method, let us consider a primitive scenario.

Company A is about to sign a treaty with two other companies, Company B and Company C. Company A considers the results of a meeting as a morphological table with two parameters—"1. B's position" and "2. C's position". Both parameters have two alternatives each—"Accept" and "Refuse". Company A also has the estimates for probability that companies B and C separately will agree to sign the treaty (Table 3.17). The specific figures are taken just for the sake of an example and bear no special significance.

With blank CCM this morphological table produces four configurations with the following probabilities:

$$
\begin{aligned}
P(\{B : accept, C : accept\}) &= 0,32;\\
P(\{B : accept, C : refuse\}) &= 0,48;\\
P(\{B : refuse, C : accept\}) &= 0,08;\\
P(\{B : refuse, C : refuse\}) &= 0,12.
\end{aligned}
\tag{3.13}
$$

These results are put into Table 3.18 for better visualization.

Let us introduce a new piece of information to this scenario: if either company refuses to sign, the other will also refuse. This can be described by the CCM in Table 3.19.

If treated using the current MMAM procedure, this table will produce the following configuration probabilities (Table 3.20).

This is clearly wrong, as configurations 2 and 3 are impossible under these restrictions. This can be mitigated by modifying CCM using derived information: if the

Table 3.18 Baseline probabilities

Scenario probability		2. Company C's position	
		2.1. Accept	2.2. Refuse
1. Company B's position	1.1. Accept	0,32	0,48
	1.2. Refuse	0,08	0,12

Table 3.19 Cross-consistency matrix 1

Cross-consistency value		2. Company C's position	
		2.1. Accept	2.2. Refuse
1. Company B's position	1.1. Accept	0	0
	1.2. Refuse	0	1

Table 3.20 Scenario probabilities 1

Scenario probability		2. Company C's position	
		2.1. Accept	2.2. Refuse
1. Company B's position	1.1. Accept	0,286	0,429
	1.2. Refuse	0,071	0,214

alternatives 1.2 and 2.2 cannot be paired with any other alternatives, then the other configurations in the respective row and column are impossible (Table 3.21).

This CCM produces a new set of configuration probabilities (Table 3.22).

This looks closer to life. One may argue that a logical step is to rewrite Table 3.21 as Table 3.23, as the configuration *{B: accept; C: accept}* is the only possible configuration if one of the alternatives, 1.1 or 2.1, is picked.

This produces a new result (Table 3.24).

Table 3.21 Cross-consistency matrix 2

Cross-consistency value		2. Company C's position	
		2.1. Accept	2.2. Refuse
1. Company B's position	1.1. Accept	0	–1
	1.2. Refuse	–1	1

Table 3.22 Scenario probabilities 2

Scenario probability		2. Company C's position	
		2.1. Accept	2.2. Refuse
1. Company B's position	1.1. Accept	0,429	0
	1.2. Refuse	0	0,571

Table 3.23 Cross-consistency matrix 3

Cross-consistency value		2. Company C's position	
		2.1. Accept	2.2. Refuse
1. Company B's position	1.1. Accept	1	–1
	1.2. Refuse	–1	1

Table 3.24 Scenario probabilities 3

Scenario probability		2. Company C's position	
		2.1. Accept	2.2. Refuse
1. Company B's position	1.1. Accept	0,727	0
	1.2. Refuse	0	0,273

This change to CCM is debatable though, as it operates on the information, which is not directly implied by the initial statement. So, for now, let us consider the results derived from Table 3.22 as official MMAM results. This raises a question though: given the known information, shouldn't the probability of *{B: refuse; C: refuse}* be at least 0,6 (the probability that C refuses)?

This scenario allows to observe the actual redistribution of probabilities under the imposed restrictions: obviously, the companies exchange their plans of accepting or refusing, and if either one refuses, the other company has to refuse regardless of their previous decision. Thus, the probability of configuration *{B: refuse; C: refuse}* is 0,2 (the probability that B refuses) plus 0,8 (the probability that B accepts) multiplied by 0,6 (the probability that C refuses); or in other words, this probability is the sum of probabilities from (3.13) where at least one company refuses because this would imply refusal by the other company due to the pre-set restriction. The realistic probabilities are presented in Table 3.25.

So, as a result of this reasoning, we can see that the configuration *{B: refuse; C: refuse}* actually gathered all of the baseline probabilities from its row and column (see Table 3.18), while the probability of *{B: accept; C: accept}* remained the same. This process may be represented from another point of view: the pair of alternatives that corresponded to the value "1" of CCM in Table 3.19 acted as a "sink" for probabilities in the respective row and column.

This concept is quite illustrious. Let us modify the initial piece of information: either company is likely to refuse to sign, if the other also refuses. This likelihood is,

Table 3.25 Scenario probabilities 4

Scenario probability		2. Company C's position	
		2.1. Accept	2.2. Refuse
1. Company B's position	1.1. Accept	0,32	0
	1.2. Refuse	0	0,68

for example, 0,5. The CCM in Table 3.19 then has the value $c_{1222} = 0{,}5$. Using the same calculation of probabilities as above, we receive the results shown in Table 3.26.

In this case, the configuration *{B: refuse; C: refuse}* again acted as a sink, but absorbed only half of the probabilities of other configurations in the row and column.

A similar line of thinking can be applied to negative CCM values, although there are important differences, which need to be taken into account. Let us consider a change in our scenario. We will replace the previous restriction for cross-consistency with the new one: both companies will not agree to sign simultaneously. This produces a new CCM (Table 3.27).

According to the previously described logic, if independently companies come to the *{B: accept; C: accept}* configuration (which should happen with 0,32 probability), one of them will change their decision, meaning that the configuration will shift either to *{B: refuse; C: accept}* or *{B: accept; C: refuse}*. Realistically both of them could change their decisions, but we will assume that the change should be minimal; a reversal in decision from either company makes the other reversal not necessary.

In this case the configuration *{B: accept; C: accept}* acts as a "source" of probability, transferring it to the other configurations in its row and column. This raises a new issue about how this surplus of probability should be distributed between the row and column. From this scenario's standpoint, the company that has a lower probability for the decision in this configuration will be more inclined to change their mind. Given the input data from Table 3.17, Company B was more likely to accept the treaty (0,8 vs. 0,4 chance), so we should distribute the configuration's probability proportionally with (1–0,8) coefficient to *{B: refuse; C: accept}* and (1–0,4) coefficient to *{B: accept; C: refuse}*. This produces the following configuration probabilities (Table 3.28).

As with the positive CCM values, the negative values in the range between –1 and 0 mean that only a portion of probability, equal to the absolute CCM value, is transferred.

Table 3.26 Scenario probabilities 5

Scenario probability		2. Company C's position	
		2.1. Accept	2.2. Refuse
1. Company B's position	1.1. Accept	0,32	0,24
	1.2. Refuse	0,04	0,4

Table 3.27 Cross-consistency matrix 4

Cross-consistency value		2. Company C's position	
		2.1. Accept	2.2. Refuse
1. Company B's position	1.1. Accept	–1	0
	1.2. Refuse	0	0

Table 3.28 Scenario probabilities 6

Scenario probability		2. Company C's position	
		2.1. Accept	2.2. Refuse
1. Company B's position	1.1. Accept	0	0,72
	1.2. Refuse	0,16	0,12

Having laid down the foundation for the proposed treatment of CCM, we can continue with the strict mathematical algorithm. To account for any possible CCM structures, several assumptions were made:

- the negative and positive CCM values are processed separately and in consequence: first, the field of configurations is considered with non-positive CCM values only as though the positive values were zero; second, the resulting field of configurations is considered with non-negative CCM values only as though the positive values were zero;
- we will call the sum of probabilities from all sources in a row (column) a "surplus" of probabilities in that row (column). This surplus is distributed between all configurations in that row (column) proportionally to both (a) the reverse CCM value, and (b) the total probability of the respective alternative in that row (column). However, the source won't accept any surplus probability if it is the most potent source in the row (column);
- if there are multiple sinks in a row (column), the power of the most potent sink becomes a taxing coefficient for that row (column). Each configuration in that row (column) gives away a portion of its probability equal to the difference between the taxing coefficient and its own CCM value. The "tax" becomes the sum of these gathered probabilities, and it becomes distributed between the configurations in that row (column) proportionally to (a) the reverse CCM value, and (b) the probability of the respective alternative in that row (column). This assumption is needed to attend to the controversial cases where the sum of CCM values is more than 1.

Using these assumptions, we can now develop an algorithm for any morphological table with two parameters, connected by a CCM of any possible structure. For simplification, we will denote $c_{1j1,2j2}$ as c_{j1j2}.

Stage 1. Calculating independent probabilities. For each configuration $\left\{a_{j1}^{(1)},\ a_{j2}^{(2)}\right\}$, the independent probability p_{j1j2} is calculated as a product of respective probabilities of alternatives:

$$p_{j1j2} = P\left(a_{j1}^{(1)}\right)P\left(a_{j2}^{(2)}\right).$$

Stage 2. Taking into account negative CCM values. For simplification, let us designate the following value:

$$c_{j1j2}^{-} = \begin{cases} c_{j1j2}, & c_{j1j2} < 0, \\ 0, & c_{j1j2} \geq 0. \end{cases}$$

Step 2.1. For each configuration $\left\{a_{j1}^{(1)}, a_{j2}^{(2)}\right\}$, calculate the surplus $s_{j1j2} = -c_{j1j2}^{-} p_{j1j2}$ and the remainder $r_{j1j2} = (1 + c_{j1j2}^{-}) p_{j1j2}$.

Step 2.2. For each row i, calculate the total row surplus:

$$s_i^h = \sum_{j=1}^{n_2} s_{ij} \frac{\left(1 - p_i^{(1)}\right)}{\left(1 - p_i^{(1)}\right) + \left(1 - p_j^{(2)}\right)}.$$

Step 2.3. For each column j, calculate the total column surplus:

$$s_j^v = \sum_{i=1}^{n_1} s_{ij} \frac{\left(1 - p_j^{(2)}\right)}{\left(1 - p_i^{(1)}\right) + \left(1 - p_j^{(2)}\right)}.$$

Step 2.4. For each configuration $\left\{a_{j1}^{(1)}, a_{j2}^{(2)}\right\}$, calculate its row distribution coefficient:

$$d_{j1j2}^h = \begin{cases} \left(\min\limits_{j2 \in [1;n_2]} c_{j1j2}^{-} - c_{j1j2}^{-}\right) p_{j1}^{(1)} / \min\limits_{j2 \in [1;n_2]} c_{j1j2}^{-}, \; s_{j1}^h > 0, \\ 0, \; s_{j1}^h = 0, \end{cases}$$

and its column distribution coefficient:

$$d_{j1j2}^v = \begin{cases} \left(\min\limits_{j1 \in [1;n_1]} c_{j1j2}^{-} - c_{j1j2}^{-}\right) p_{j2}^{(2)} / \min\limits_{j1 \in [1;n_1]} c_{j1j2}^{-}, \; s_{j2}^v > 0, \\ 0, \; s_{j2}^v = 0. \end{cases}$$

Step 2.5. For each configuration $\{a_{j1}^{(1)}, a_{j2}^{(2)}\}$, calculate its normalized row distribution coefficient:

$$\tilde{d}_{j1j2}^h = \begin{cases} d_{j1j2}^h / \sum_{j2=1}^{n_2} d_{j1j2}^h, \; s_{j1}^h > 0, \\ 0, \; s_{j1}^h = 0, \end{cases}$$

and its normalized column distribution coefficient:

$$\tilde{d}^v_{j1j2} = \begin{cases} d^v_{j1j2}/\sum_{j1=1}^{n_1} d^v_{j1j2}, & s^v_{j2} > 0, \\ 0, & s^v_{j2} = 0. \end{cases}$$

Step 2.6. For each configuration $\left\{a^{(1)}_{j1},\ a^{(2)}_{j2}\right\}$, calculate the new probability:

$$p'_{j1j2} = r_{j1j2} + \tilde{d}^h_{j1j2} s^h_{j1} + \tilde{d}^v_{j1j2} s^v_{j2}.$$

Stage 3. Taking into account positive CCM values. For simplification, let us designate the following value:

$$c^+_{j1j2} = \begin{cases} c_{j1j2}, & c_{j1j2} > 0, \\ 0, & c_{j1j2} \le 0. \end{cases}$$

Step 3.1. For each row i, determine the taxing coefficient: $T^h_i = \max_{j \in [1;n_2]} c^+_{ij}$.

Step 3.2. For each column j, determine the taxing coefficient: $T^v_j = \max_{i \in [1;n_1]} c^+_{ij}$.

Step 3.3. For each configuration $\left\{a^{(1)}_{j1},\ a^{(2)}_{j2}\right\}$, calculate the tax:

$$t_{j1j2} = \max\left(T^h_{j1} - c^+_{j1j2},\ T^v_{j2} - c^+_{j1j2}\right) p'_{j1j2},$$

and the remainder:

$$r'_{j1j2} = \left(1 - \max\left(T^h_{j1} - c^+_{j1j2},\ T^v_{j2} - c^+_{j1j2}\right)\right) p'_{j1j2}.$$

Step 3.4. For each row i, calculate the total row tax:

$$t^h_i = \sum_{j=1}^{n_2} t_{ij}.$$

Step 3.5. For each column j, calculate the total column tax:

$$t^v_j = \sum_{i=1}^{n_1} t_{ij}.$$

Step 3.6. For each configuration $\left\{a^{(1)}_{j1},\ a^{(2)}_{j2}\right\}$, calculate its row distribution coefficient:

$$d'^h_{j1j2} = \begin{cases} c^+_{j1j2} p^{(1)}_{j1} / T^h_{j2}, & t^h_{j1} > 0, \\ 0, & t^h_{j1} = 0, \end{cases}$$

and its column distribution coefficient:

$$d'^{v}_{j1j2} = \begin{cases} c^{+}_{j1j2} p^{(2)}_{j2} / T^{v}_{j1}, \ t^{v}_{j2} > 0, \\ 0, \ t^{v}_{j2} = 0. \end{cases}$$

Step 3.7. For each configuration $\left\{a^{(1)}_{j1}, a^{(2)}_{j2}\right\}$, calculate its normalized row distribution coefficient:

$$\tilde{d}'^{h}_{j1j2} = \begin{cases} d^{h}_{j1j2} / \sum_{j2=1}^{n_2} d^{h}_{j1j2}, \ s^{h}_{j1} > 0, \\ 0, \ s^{h}_{j1} = 0, \end{cases}$$

and its normalized column distribution coefficient:

$$\tilde{d}'^{v}_{j1j2} = \begin{cases} d^{v}_{j1j2} / \sum_{j1=1}^{n_1} d^{v}_{j1j2}, \ s^{v}_{j2} > 0, \\ 0, \ s^{v}_{j2} = 0. \end{cases}$$

Step 3.8. For each configuration $\left\{a^{(1)}_{j1}, a^{(2)}_{j2}\right\}$, calculate the new probability:

$$p''_{j1j2} = r'_{j1j2} + \tilde{d}'^{h}_{j1j2} t^{h}_{j1} + \tilde{d}'^{v}_{j1j2} t^{v}_{j2}.$$

The obtained p''_{j1j2} are the configuration probability values that take into account the inter-dependence between parameters. As with the usual MMAM procedure, the new alternative probabilities can be calculated by gathering the individual configuration probabilities bearing that alternative:

$$p''^{(1)}_{j1} = \sum_{j2=1}^{n_2} p''_{j1j2}.$$

Generalization of the algorithm. While the developed algorithm is most presentable for a two-parameter morphological table, it can be generalized for morphological tables with an arbitrary number of parameters. The tricky part of this procedure is that each configuration now has a corresponding set of CCM values for each pair of alternatives that it contains, so a configuration can act simultaneously as a source and as a sink depending on the chosen two-dimensional slice of the multidimensional configuration cube.

Thus, the following upgrades on the presented procedure are made:

- at Stage 2, any configuration is considered a source if it has at least one negative CCM value in its set. This configuration provides a surplus of probability, with its portion determined by the minimum CCM value among all that correspond to this configuration. This surplus is distributed among all of the parameters proportionally to the product of minimum CCM value among all that correspond to this parameter in the configuration, and one minus probability of the alternative for this parameter in the configuration. The rest of Stage 2 proceeds as described above, except that row/column should be replaced by *i*th parameter;

- at Stage 3, each configuration is considered in the axis of each parameter, and the probability tax is determined by the most potent sink in either of the axes. The rest of Stage 3 proceeds as described above, except that row/column should be replaced by *i*th parameter.

The presented approach is a promising method of treating CCM in MMAM, as it provides a meaningful way of understanding and taking into account CCM values. However, it should be noted that the basic scenario that was used as a demonstration, was crafted specifically so that the processes of transferring probability could be observed. In reality, most of the problems do not have evident ways of forming probabilities, so the reasoning which led to create this approach, may be invalid.

It is easy to construct a scenario that would be a counterexample: let us imagine a simple morphological table for traffic accidents with two parameters: (1) place of the accident (crossroads/crossing, or middle of the road), and (2) cause of the accident (speeding, or movement at red light). It is obvious that "movement at red light" alternative should always be paired with "crossroads/crossing" alternative, but as the inverse statement is not true, no possible CCM can cover this scenario correctly.

The main limitation is the obligatory symmetry assumption of the CCM in both older and newer approaches. Breaking this symmetry, while possible mathematically, leads to other organizational issues related to a sudden increase in complexity for experts and analyst.

Thus, the choice of CCM processing method lies on the analyst, and different approaches might be suitable for different problems. The ideal approach remains to be developed.

3.6 Two-Stage Modified Morphological Analysis Method

The scenario analysis process often makes useful the application of the two-stage MMAM procedure. It provides the study of uncontrolled, or external world, factors at the first stage, and assesses the decision parameters at the second stage, taking into account the whole multitude of potential configurations defined by the MT at the first stage (Fig. 3.2).

Here two related morphological tables are formed. The first stage MT is called the scenario morphological table, and the second stage MT is called the strategy morphological table.

The specifics of the second stage of MMAM is that the selection of alternatives of the strategy MT depends on the decision-maker, instead of some random uncontrolled factors, therefore, it does not make sense to consider their probabilities. To assess the alternatives and configurations in the second stage, a value which we call expected

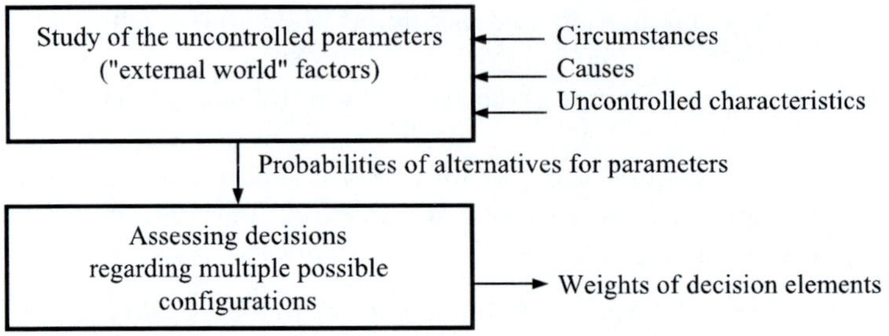

Fig. 3.2 Two-stage MMAM scheme

efficiency is used, i.e., the likelihood that the selection of the respective alternative or configuration will lead to desirable results.

Dependency matrix. The efficiency of the parameter alternatives in the strategy MT depends on the external data, in this case, the parameters in the scenario MT. To process these dependencies, a dependency matrix is proposed, which is quite similar to the CCM matrix used in the first stage of the MMAM study. However, in the case of a dependency matrix, the relation between the alternatives is one-sided.

According to the developed procedure, each pair of alternatives $a_{j_1}^{(i_1)}$, $a_{j_2}^{(i_2)}$ of the parameters F_{i_1}, F_{i_2} of MTs from the first and second stages, respectively, is assigned a value $c_{i_1 j_1, i_2 j_2} \in [-1; 1]$ according to Table 3.29.

The dependency matrix is filled by the experts via a procedure similar to filling the CCM matrix (as described in Sect. 3.4).

The problem that appears in the second stage of morphological study, can be stated as such:

Given:

- the scenario MT with N parameters and the strategy MT with N' parameters. The tables contain a combined set of characteristic parameters $F = \{F_i | i \in$

Table 3.29 Dependency matrix values' explanation

Value	Explanation
−1	The alternative of the strategy MT parameter is absolutely ineffective when choosing the respective alternative of the scenario MT
(−1;0)	Appearance of the alternative of the scenario MT reduces the efficiency of the alternative of the strategy MT
0	Appearance of the alternative of the scenario MT has no impact on the efficiency of the alternative of the strategy MT
(0;1)	Appearance of the alternative of the scenario MT increases the efficiency of the alternative of the strategy MT
1	The alternative of the strategy MT parameter is absolutely effective when choosing the respective alternative of the scenario MT

$\overline{1,\ N+N'}\}$, with each parameter F_i described by a set of alternatives $A_i = \left\{a_j^{(i)} \middle| j \in \overline{1,\ n_i}\right\}$;

- the probabilities of alternatives from the scenario MT $p_j^{(i)},\ i \in \overline{1, N},\ j \in \overline{1,\ n_i}$ (calculated at the first MMAM stage);
- the values of the dependency matrix for each pair of alternatives, where the first alternative is from the scenario MT, and the second alternative is from the strategy MT $\left\{c_{i_1 j_1, i_2 j_2} | i_1 \in \overline{1,\ N},\ i_2 \in \overline{N+1,\ N+N'};\ \ j_1 \in \overline{1,\ n_{i_1}};\ j_2 \in \overline{1,\ n_{i_2}}\right\}$.

Required:

- to calculate the expected efficiencies $R_j^{(i)}$ of each of the alternatives $a_j^{(i)},\ i \in \overline{N+1, N+N'},\ j \in \overline{1,\ n_i}$ from the strategy MT, considering that any possible configuration of the scenario MT may emerge;
- to calculate the expected efficiencies $R\{s_l\}$ of configurations s_l, defined by the strategy MT considering that any possible configuration of the scenario MT may emerge.

The choice of an alternative for a strategy MT parameter is a decision under conditions of different circumstances that may appear with a certain probability. To calculate the expected efficiency, we should consider all possible configurations from the scenario MT, analyzing the efficiency of an alternative in conditions of each of the scenario MT configurations.

Let us introduce a value similar to (3.7) which we call the conditional efficiency $R\left(a_j^{(i)} \middle| \left\{a_{j_1}^{(1)}, a_{j_2}^{(2)},\ a_{j_3}^{(3)},\ \ldots,\ a_{j_N}^{(N)}\right\}\right)$ for an alternative $a_j^{(i)},\ i \in \overline{N+1, N+N'}$ when considered under the circumstances of the scenario MT configuration $\left\{a_{j_1}^{(1)}, a_{j_2}^{(2)},\ a_{j_3}^{(3)},\ \ldots,\ a_{j_N}^{(N)}\right\}$:

$$R\left(a_j^{(i)} \middle| \left\{a_{j_1}^{(1)}, a_{j_2}^{(2)},\ a_{j_3}^{(3)},\ \ldots,\ a_{j_N}^{(N)}\right\}\right) = \frac{p_j'^{(i)} \cdot \prod_{m=1}^{N} \left(c_{mj_m,ij} + 1\right)}{\sum_{k=1}^{n_i} p_k'^{(i)} \cdot \prod_{m=1}^{N} \left(c_{mj_m,ik} + 1\right)},$$

where $p_{j_i}'^{i}$ is the preliminary efficiency assessment for the alternative $a_{j_i}^{(i)}$. In most cases there is no additional data about this preliminary efficiency, so for each of the parameters all of its alternatives are given the same value preliminary efficiency value $p_j'^{(i)} = 1/n_i$.

This is a case of decision-making under uncertainty conditions, so the expected efficiency of the alternative $a_j^{(i)},\ i \in \overline{N+1, N+N'}$ may be obtained using the following expression (Zaychenko 2006):

$$R_j^{(i)} = \sum_{j_1=1}^{n_1} \sum_{j_2=1}^{n_2} \ldots \sum_{j_N=1}^{n_N} R\left(a_j^{(i)} \middle| \left\{a_{j_1}^{(1)}, a_{j_2}^{(2)},\ a_{j_3}^{(3)},\ \ldots,\ a_{j_N}^{(N)}\right\}\right) P\left(\left\{a_{j_1}^{(1)}, a_{j_2}^{(2)},\ a_{j_3}^{(3)},\ \ldots,\ a_{j_N}^{(N)}\right\}\right),$$

where $P\left(\left\{a_{j_1}^{(1)}, a_{j_2}^{(2)}, a_{j_3}^{(3)}, \ldots, a_{j_N}^{(N)}\right\}\right)$ is the configuration $\left\{a_{j_1}^{(1)}, a_{j_2}^{(2)}, a_{j_3}^{(3)}, \ldots, a_{j_N}^{(N)}\right\}$ probability, calculated at the first stage of the MMAM study.

The produced values are convenient for ranking the alternatives of the strategy MT parameters by their expected efficiency.

An alternative assessment $W_j^{(i)}$ of the expected efficiency for the alternatives of the strategy MT is to calculate the distance to the hypothetical "ideal" strategy, which selects the most favorable alternatives from the strategy MT for each of the possible scenario MT configurations:

$$W_j^{(i)} = \sum_{j_1=1}^{n_1} \sum_{j_2=1}^{n_2} \cdots \sum_{j_N=1}^{n_N} (R(a_{\max_i}^{(i)} |\{a_{j_1}^{(1)}, a_{j_2}^{(2)}, a_{j_3}^{(3)}, \ldots, a_{j_N}^{(N)}\})-$$
$$-R(a_j^{(i)} |\{a_{j_1}^{(1)}, a_{j_2}^{(2)}, a_{j_3}^{(3)}, \ldots, a_{j_N}^{(N)}\})) \cdot P(\{a_{j_1}^{(1)}, a_{j_2}^{(2)}, a_{j_3}^{(3)}, \ldots, a_{j_N}^{(N)}\}) = R_{\max}^{(i)} - R_j^{(i)},$$

where $a_{\max_i}^{(i)}$ is the alternative having maximum $R\left(a_j^{(i)}\middle|\left\{a_{j_1}^{(1)}, a_{j_2}^{(2)}, a_{j_3}^{(3)}, \ldots, a_{j_N}^{(N)}\right\}\right)$ value, and

$$R_{\max}^{(i)} = R\left(a_{\max_i}^{(i)}\middle|\left\{a_{j_1}^{(1)}, a_{j_2}^{(2)}, a_{j_3}^{(3)}, \ldots, a_{j_N}^{(N)}\right\}\right) P\left(\left\{a_{j_1}^{(1)}, a_{j_2}^{(2)}, a_{j_3}^{(3)}, \ldots, a_{j_N}^{(N)}\right\}\right)$$

is the efficiency of the hypothetical "ideal" strategy that chooses the most efficient alternative from the strategy MT under conditions of each possible configuration from the scenario MT.

The value $W_j^{(i)}$ is also useful, as it shows the expected fall of the efficiency for the alternative $a_j^{(i)}$ considering that the unfavorable configurations of circumstances are possible. The value $W_j^{(i)} = 0$ only when the alternative $a_j^{(i)}$ is the most efficient in all possible configurations.

3.7 Constructing and Processing Networks of Morphological Tables

Complex problems in scenario analysis usually involve systems with a number of interacting entities, most of which have inherent uncertainties, so the multitude of possible configurations for these entities should be considered, and a separate morphological table should be constructed for each of the entities. The entities themselves can have very different natures: they may represent material objects or systems, as well as events or processes, and even abstract concepts (e.g., the consequences of a catastrophe).

To deal with such problems, the morphological tables are combined into a network and are processed by a generalization of the two-stage MMAM procedure. The morphological tables in a network have three possible types of relations between the tables:

1. *bi-directional*. This type of relation means that the objects are interdependent, and determining their probabilistic state happens simultaneously. A relation of this type is defined by a cross-consistency matrix. Technically, from the method's point of view, a group of tables linked by bi-directional relations creates a single combined morphological table, that is processed by a unified system of equations. However, a modular approach has many advantages. As each table in a group has its own cross-consistency matrix, other tables can be added or removed from the group without having to re-evaluate the whole virtual cross-consistency matrix uniting all tables in a group. Single morphological tables in a group can be assessed by different groups of experts, that have knowledge exactly in the field of the corresponding morphological table. A repository of these tables with pre-estimated cross-consistency matrices can be created if the problem is going to be analyzed several times. As for the consistency matrices that connect tables in a group, they are mostly sparse and easier to estimate.
2. *unidirectional*. The unidirectional, or causal type of relation is the most common for networks of morphological tables, when the state of the second, dependent table is influenced by the state of the first table. This type of relation corresponds to the two-stage morphological analysis procedure, where the independent parameters are calculated at the first stage, and then the expected efficiencies of the second morphological table are calculated at the second stage.
3. *hierarchical*. Large networks of tables may require creating aggregated tables that generalize the state of lower-level morphological tables. The alternatives of a parameter in an aggregated table have probabilities that correspond to the sum of probabilities for groups of configurations at lower levels of the hierarchy.

The dependencies between morphological tables can be presented as a graph. A single table constitutes a node of the graph, and a unidirectional link represents an arc. When all the necessary data has been obtained (i.e., the preliminary assessments for alternatives, the values of CCMs and dependency matrices), the graph processing procedure continues as follows:

1. Each cycle in a graph is aggregated into a single node. The morphological tables for nodes in a cycle are combined into a virtual large morphological table. The CCM for this table is formed by uniting the respective CCMs for constituent tables. After that, an acyclic graph is obtained.
2. For each of the morphological tables that corresponds to a source of the graph, a one-stage MMAM procedure is conducted to take into account the interdependencies between alternatives.
3. For each node of the graph that depends directly only on the calculated nodes, a dependency matrix is processed as if it was a two-stage morphological procedure (see Sect. 3.6). The obtained values become the preliminary alternative assessments in a morphological table at the dependent node.
4. For each node, where all the dependencies from influencing nodes are processed, a one-stage MMAM procedure is performed.
5. Steps 3 and 4 are repeated until all the nodes of the graph become calculated.

Typically, the sinks of the graph correspond to the morphological tables for the decisions.

3.8 Software Implementation of the Modified Morphological Analysis Method

Most of the described MMAM procedures were implemented using the SAS Studio software with user C# modules, compiled in the Microsoft Visual Studio environment. The modules correspond to the main steps of MMAM: MT construction; MT estimation; cross-consistency matrix estimation; and weight calculation for one-stage and two-stage MMAM procedures. This tool set allows to design and perform morphological studies for complex tasks.

Let us overview the capacities of the modules.

The morphological table construction module is shown in Fig. 3.3.

This module allows to form a list of morphological table parameters and designate the alternatives for each of them. The result contains the records with the parameter and alternative identifiers, parameter and alternative names, and affiliation of the alternatives to the parameters.

Morphological table evaluation module. This module requires an input in the form of a morphological table, created by the previous module. The assessments may be made using the combo box with Miller's scale descriptions, as well as be entered manually. The main module window is shown in Fig. 3.4.

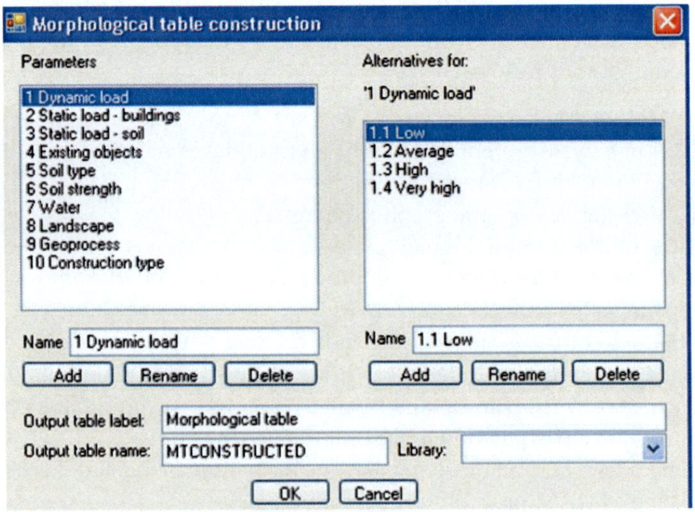

Fig. 3.3 Morphological table construction module

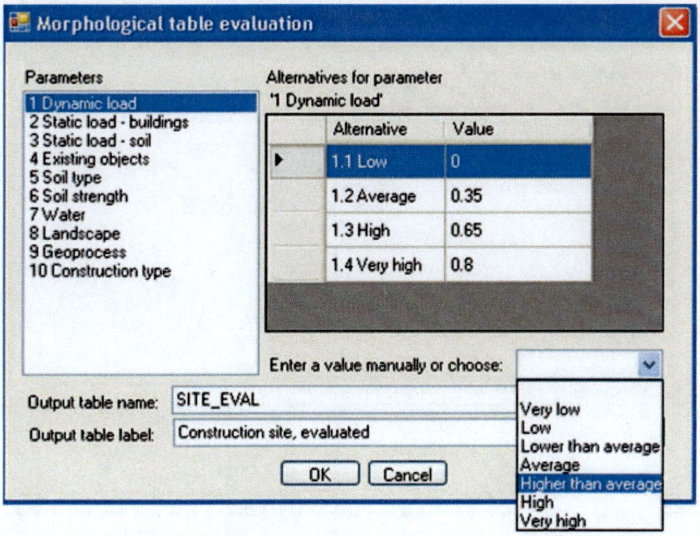

Fig. 3.4 Morphological table evaluation module

As a result of this module's work, a table is created that contains alternative identifiers and their respective values. It should be noted that a single constructed morphological table can have multiple evaluations by adding the duplicates of this module with varying estimates.

Cross-consistency matrix evaluation module. This module also requires an input in the form of a morphological table, created by the morphological table construction module. It displays a fragment of the CCM for the selected pair of parameters. As with the previous module, the CCM value can be assigned manually or chosen from a list in a combo box with verbal descriptions. The main window is shown in Fig. 3.5:

As a result of this module's work, a table is created that contains two alternative identifiers and the corresponding CCM value.

One-stage morphological analysis calculation module. This module receives an input in the form of the morphological table description, preliminary assessment values, and CCM values; however, only the morphological table description is obligatory. The module conducts the one-stage MMAM procedure, its main window is presented in Fig. 3.6.

The output of this table is a table with estimates of the alternatives. It is identical in form to the morphological table evaluation module output, but these estimates take into account the interdependencies between parameters.

Result output module. This module receives the morphological table description, and any number of evaluations for this morphological table, and displays them in a presentable form. The data may be output both as a SAS report and as an html file that can be downloaded. The module's main window is shown in Fig. 3.7.

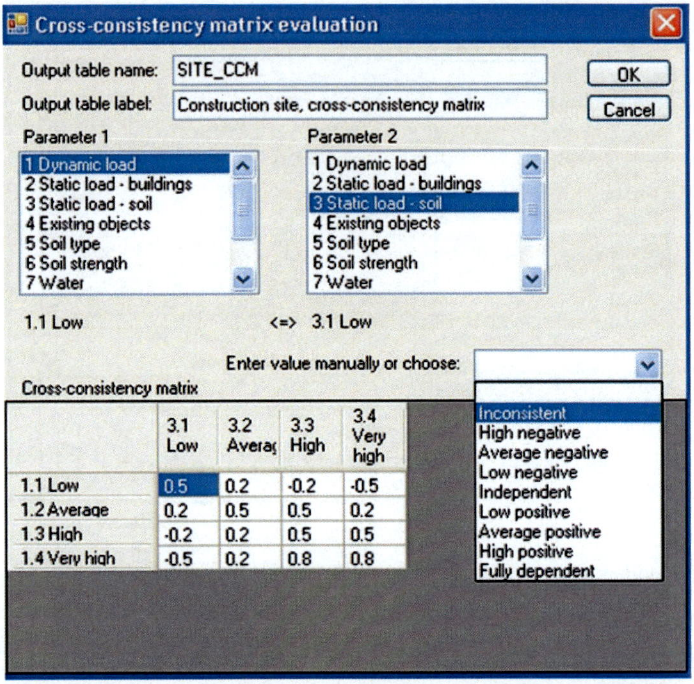

Fig. 3.5 Cross-consistency matrix evaluation module

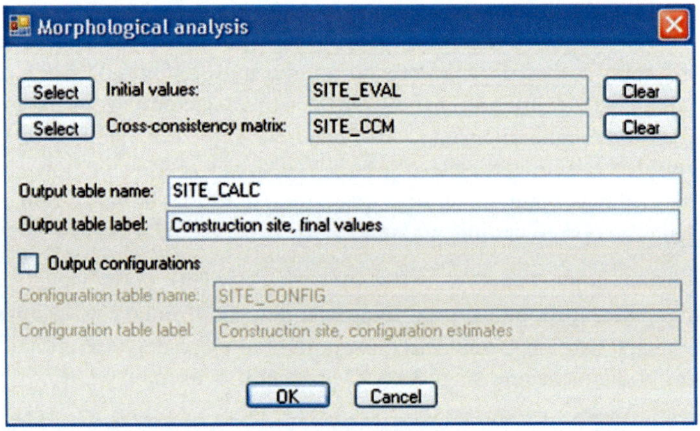

Fig. 3.6 One-stage morphological analysis calculation module

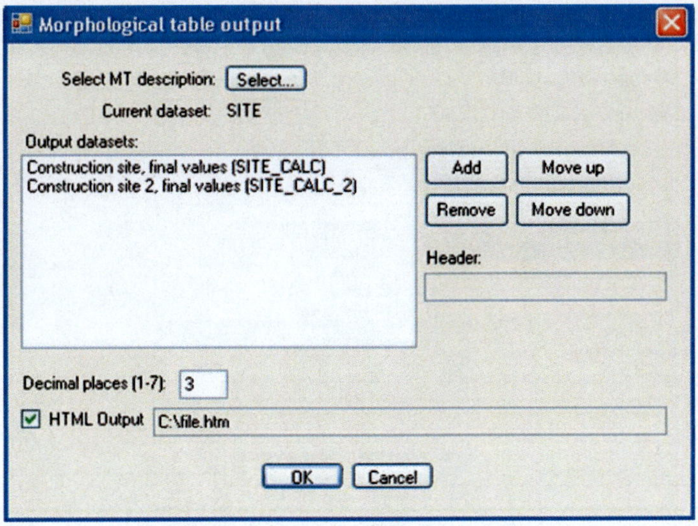

Fig. 3.7 Result output module

A fragment of the result of this module's work is shown in Fig. 3.8: it displays the morphological table with the respective estimates for the alternatives. Several different evaluation sets may be displayed for comparison, for example, the preliminary and final values, or the resulting values for different objects.

<u>Dependency matrix evaluation module.</u> The module is similar to the cross-consistency matrix, except it requires two different morphological table descriptions

Construction site, final values

1 Dynamic load		2 Static load - buildings		3 Static load - soil		4 Existing objects		5 Soil type		6 Soil strength		7 Water		8 Land
Alternative	Weight	Alternative	Weight	Alternative	Weight	Alternative	Weight	Alternative	Weight	Alternative	Weight	Alternative	Weight	Alternativ
1.1 Low	0.000	2.1 Low	0.000	3.1 Low	0.030	4.1 None	0.000	5.1	0.107	6.1 Very strong	0.000	7.1 Far	0.000	8.1 Flat
1.2 Average	0.388	2.2 Average	0.000	3.2 Average	0.717	4.2 Far	0.479	5.2	0.372	6.2 Strong	0.000	7.2 >3 >10	0.384	8.2 1-4 degrees
1.3 High	0.286	2.3 High	0.000	3.3 High	0.189	4.3 Average	0.521	5.3	0.434	6.3 Average	0.352	7.3 <3 >10	0.552	8.3 4-8 degrees
1.4 Very high	0.326	2.4 Very high	1.000	3.4 Very high	0.063	4.4 Near	0.000	5.4	0.087	6.4 Weak	0.648	7.4 Near	0.065	8.4 >8 degrees

Page Break

Construction site 2, final values

1 Dynamic load		2 Static load - buildings		3 Static load - soil		4 Existing objects		5 Soil type		6 Soil strength		7 Water		8 Land
Alternative	Weight	Alternative	Weight	Alternative	Weight	Alternative	Weight	Alternative	Weight	Alternative	Weight	Alternative	Weight	Alternativ
1.1 Low	0.707	2.1 Low	0.000	3.1 Low	0.324	4.1 None	0.714	5.1	0.170	6.1 Very strong	0.000	7.1 Far	0.098	8.1 Flat
1.2 Average	0.263	2.2 Average	1.000	3.2 Average	0.639	4.2 Far	0.286	5.2	0.450	6.2 Strong	0.209	7.2 >3 >10	0.672	8.2 1-4 degrees
1.3 High	0.029	2.3 High	0.000	3.3 High	0.025	4.3 Average	0.000	5.3	0.342	6.3 Average	0.670	7.3 <3 >10	0.126	8.3 4-8 degrees
1.4 Very high	0.000	2.4 Very high	0.000	3.4 Very high	0.012	4.4 Near	0.000	5.4	0.038	6.4 Weak	0.120	7.4 Near	0.104	8.4 >8 degrees

Page Break

Fig. 3.8 Calculated values of alternatives, as presented by the result output module

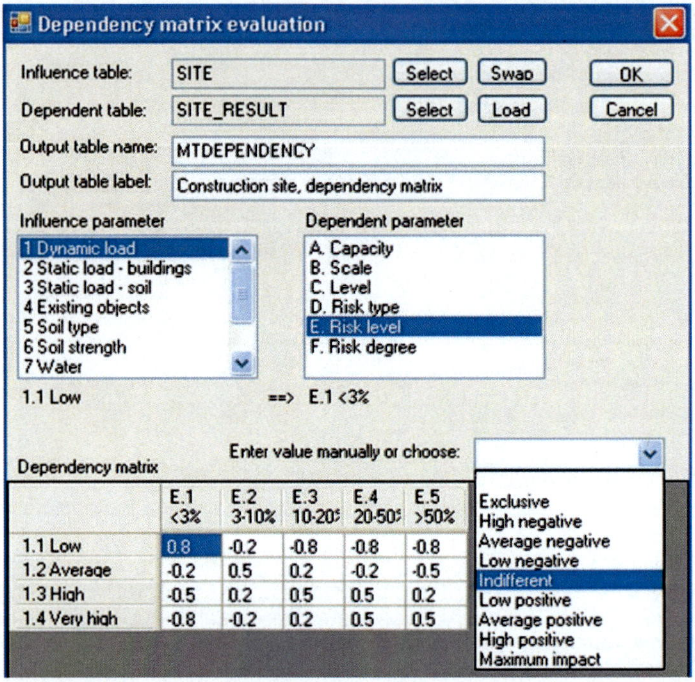

Fig. 3.9 Dependency matrix evaluation module

as an input. Accordingly, to display and fill in the dependency matrix, two parameters from different tables should be selected. The module's main window is shown in Fig. 3.9.

Two-stage morphological analysis calculation module. The module implements the two-stage MMAM calculation procedure (see Sect. 3.6). It requires as an input: the descriptions of the first and second morphological tables, the preliminary assessments for their alternatives, the CCM for the first stage, and the dependency matrix. The module's main window, where all the input records are determined, is shown in Fig. 3.10.

The resulting table is a standard evaluation record for the dependent morphological table. The estimates may be viewed using the previously described output module.

The developed modules may be used for complex MMAM problems. A typical two-stage procedure with a comparison of two objects is shown in Fig. 3.11.

The created tool set is flexible and powerful, allowing to model multi-factor problems and support decision-making for complex systems of different natures.

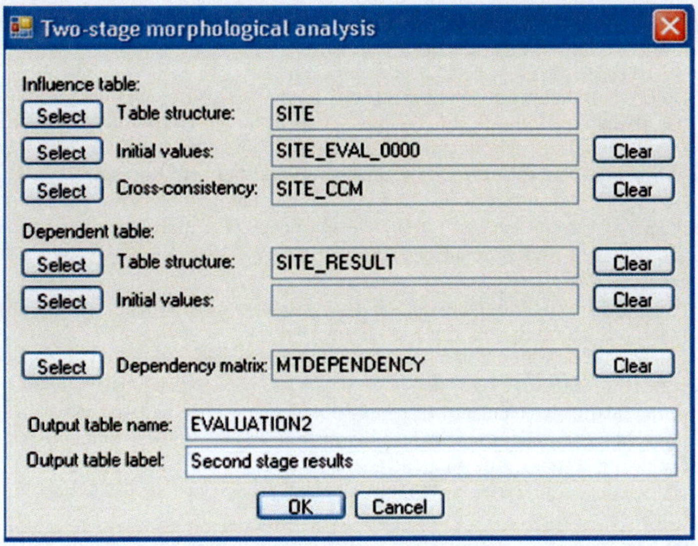

Fig. 3.10 Two-stage morphological analysis calculation module

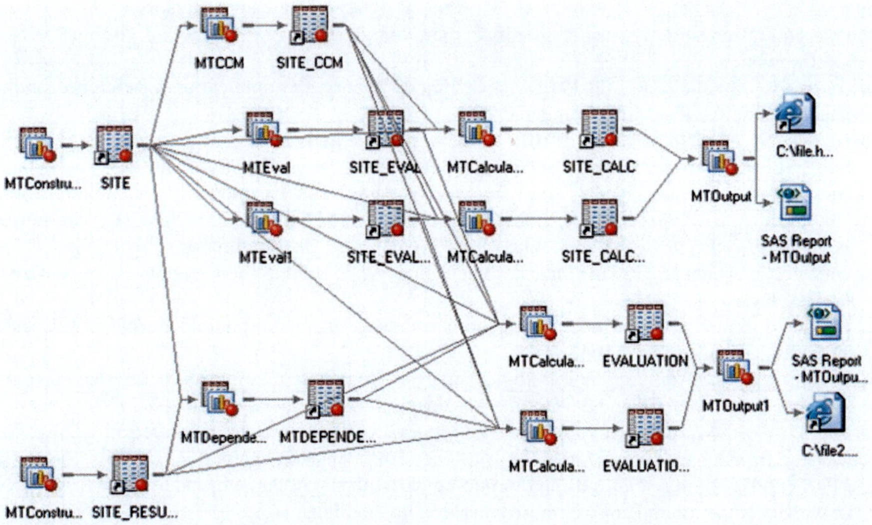

Fig. 3.11 A sample morphological model in the developed environment

References

Akimov S (2001) morphological analysis of uhf linear transistor amplifiers. Communication Educational Institutions Reports 166, 84–89, SPbGUT, Saint Petersburg (In Russian)

Akimov S (2005) Model of identification level morphological set. Communication Educational Institutions Reports 172, 120–135, SPbGUT, Saint Petersburg (In Russian)

Ayres R (1969) Technological forecasting and long-range planning. McGraw-Hill Book Company

Haiko H, Savchenko I, Matviichuk I (2019) Development of a morphological model for territorial development of underground city space. Naukovyi Visnyk Natsionalnoho Hirnychoho Universytetu 3:92–98

Hussain N, Ritchey T (2011) Achieving clarity in complex pharmaceutical problems: from innovation to policy formulation. Eur Ind Pharm 10:4–7

Isaksson S, Ritchey T (2003) Protection against sabotage of nuclear facilities: using morphological analysis in revising the design basis threat. In: Adaptation of a paper delivered to the 44th annual meeting of the Institute of Nuclear Materials Management, Phoenix, Arizona

Levin M, Danieli M (2005) Hierarchical decision making framework for evaluation and improvement of composite systems. Informatica 16(2):213–240

Levin M, Vishnitskiy R (2007) Towards morphological design of GSM network. Informational Processes 7(2):183–190

Odrin V (1986) Morphological synthesis of systems: statement of the problem, classification of methods, morphological construction methods. V.M. Glushkov Institute of Cybernetics, Kyiv (In Russian)

Odrin V, Kartavov S (1977) Morphological analysis of systems. Naukova Dumka, Kyiv (In Russian)

Pankratova N, Savchenko I, Gayko G, Kravets V (2018) Evaluating perspectives of urban underground construction using modified morphological analysis method. J Autom Inf Sci 50(10):34–46

Pankratova N, Malafieieva L (2017) Delphi method. Theory and applications. Reference book. Naukova Dumka, Kyiv (In Ukrainian)

Pankratova N, Nedashkivska N (2010) Models and methods of hierarchy analysis. Theory. Applications: reference book. "Polytechnica" Publishing, Kyiv (In Ukrainian)

Pankratova N, Savchenko I (2008) Applying morphological analysis method to technology foresight problems. Reports of Petro Mohyla Mykolayiv State Humanitarian University. Computer technology, system analysis, modeling series 77(90), 6–13 (In Ukrainian)

Pankratova N, Savchenko I (2015) Morphological analysis. Problems, theory, application. Naukova Dumka, Kyiv

Petrusel R, Mocean L (2007) Modeling decisional situations using morphological analysis. Revista Informatica Economic 4(44):41–44

Pluzhnikov A (1987) Applying morphological analysis for search of rational structures and layouts of mechanisms. Method and development. Minzhivmash, Lyubertsy (In Russian)

Ritchey T (1998) Morphological analysis – a general method for non-quantified modeling. In: Adapted from a paper presented at the 16th Euro conference on operational analysis, Brussels

Ritchey T (2005) Futures studies using morphological analysis. Adapted from an article for the UN University Millennium Project: futures research methodology series

Ritchey T (2006) Problem structuring using computer-aided morphological analysis. J Oper Res Soc 57:792–801

Ritchey T (2009) Threat analysis for the transport of radioactive material. Packag Transp Storage Secur Radioactive Mater 20(1)

Ritchey T (2011) Modeling alternative futures with general morphological analysis. World Future Rev, 83–94

Ritchey T, Zwicky F (1998) "Morphologie" and policy analysis. In: Adapted from an paper presented at the 16th EURO Conference on Operational Analysis, Brussels

Saaty T (1993) Decision making. Analytical hierarchy process. Radio and communication, Moscow (In Russian)

Savchenko I (2011) Methodological and mathematical support for solving foresight problems using modified morphological analysis method. Syst Res Inf Technol 3:18–28 (In Ukrainian)

Savchenko I (2016) Estimating the solution sensitivity in application of the modified morphological analysis method. Cybern Syst Anal 52(5):782–790

Savchenko I (2015) Methodological and mathematical support for solving foresight problems using modified morphological analysis method. In: Gorelova G, Pankratova N (eds) IInnovative development of socioeconomic systems based on foresight and cognitive modeling methodologies. Naukova Dumka, Kyiv, pp 427–441 (In Russian)

Savchenko I (2018) Using morphological table networks for modeling social disaster situations. In: 2018 IEEE first international conference on system analysis & intelligent computing, Kyiv

Savchenko I (2022) Detecting and handling flawed input data in modified morphological analysis method. In: 2022 IEEE 3rd international conference on system analysis & intelligent computing, Kyiv

Zaychenko Yu (2006) Operation research: textbook, 7th edn. Publishing House "Slovo", Kyiv

Zgurovsky M, Pankratova N (2005) Technology foresight. "Polytechnica" Publishing, Kyiv (In Ukrainian)

Zgurovsky M, Pankratova N (2007) System analysis: theory and applications. Springer

Zwicky F (1968) Discovery, invention, research through the morphological approach. The Macmillan Co

Zwicky F, Wilson A (1967) New methods of thought and procedure. In: Contributions to the symposium on methodologies, Pasadena, pp 273−297

Chapter 4
Strategy of Evaluation and Risk Management in Practical Urban Development Problems

Abstract A number of application cases of the modified morphological analysis method in practical urban development problems are presented. The technique for studying detrimental events, such as traffic accidents and jams, using the modified morphological analysis method is given, providing a well-grounded decision support regarding the mitigation methods for these events. A morphological model allowing to assess sites for underground construction is considered, where an uncertain or heterogeneous geological environment is studied at the pre-project stage using the modified morphological analysis method to obtain the estimates for its general favorability for construction, and the related recommended characteristics and involved risks of various types. This technique is complemented by a morphological model for structural and functional factors, providing an opportunity to evaluate specific construction objects—in the presented cases, parking lots and car tunnels. Another case of the method's application is the comparison of different types of urban objects (underwater drainage pipes to underwater drainage tunnel; bridge crossing to tunnel crossing) regarding their resilience to undesirable events, including those of natural, technogenic, or anthropogenic origin. Finally, a framework for using a network of morphological tables for preventing and mitigating disasters is presented.

4.1 Traffic Accident Scenarios Evaluation

An important problem for the normal functioning of the large city's traffic system is the spontaneous occurrence of accidents that paralyze or hinder movement for a period of time on some routes. To study the regularities of accidents, a morphological research was made (Pankratova 2009), that allowed also to assess methods of countering accidents and justify constructing car tunnels for Kyiv city.

The morphological study incorporated the following parameters that describe a traffic accident:

1. type of the accident: collision with a stationary object; collision of two or more vehicles; running-down accident; vehicle failure;

N. Pankratova et al., *Modeling the Underground Infrastructure of Urban Environments*, The Urban Book Series, https://doi.org/10.1007/978-3-031-47522-1_4

2. place of the accident: crossing; small road; wide road; tunnel; bridge; parking lot; yard;
3. time of the accident: day (natural illumination); night (artificial illumination);
4. weather conditions: normal; rain; snow; icy; fog;
5. driver's condition: normal; alcoholic intoxication;
6. trigger of the accident: excessive speed; traffic area violation; movement under red light; illegal turn; illegal or unexpected stop; movement in illegal area; independent of the vehicle factors;
7. cause of the accident: driver's carelessness or error; loss of control; conscious violation of traffic rules; vehicle malfunction; driver-independent factors.

Graphic representation of this set of parameters and alternatives is given in Table 3.2.

Preliminary probability values were estimated by the experts using pair-wise comparison. The questions about the pairs of alternatives were stated in the form "Which of the alternatives of an accident's parameter happens more often? How high is this preference?" The obtained pair-wise comparison matrices are shown in Tables 4.1, 4.2, 4.3, 4.4, 4.5, 4.6 and 4.7.

After processing the pair-wise comparison matrices the weights of the alternatives are obtained, corresponding to the preliminary (independent) probability estimates (Table 4.8).

The cross-consistency matrix for describing interdependency between the parameters is also assessed (Table 4.9). Empty cells denote zero; as the matrix is symmetrical, only its half is shown.

Table 4.1 Pair-wise comparison matrix for parameter 1 "Type of the accident"

	1.1	1.2	1.3	1.4	Estimate
1.1	1	1/3	2	4	0,273
1.2	3	1	4	5	0,484
1.3	1/2	1/4	1	3	0,177
1.4	1/4	1/5	1/3	1	0,066

Table 4.2 Pair-wise comparison matrix for parameter 2 "Place of the accident"

	2.1	2.2	2.3	2.4	2.5	2.6	2.7	Estimate
2.1	1	3	2	5	6	5	5	0,311
2.2	1/3	1	1/2	3	4	3	2	0,159
2.3	1/2	2	1	5	6	5	5	0,282
2.4	1/5	1/3	1/5	1	2	1	2	0,078
2.5	1/6	1/4	1/6	1/2	1	1/2	1/2	0,036
2.6	1/5	1/3	1/5	1	2	1	2	0,078
2.7	1/5	1/2	1/5	1/2	2	1/2	1	0,056

Table 4.3 Pair-wise comparison matrix for parameter 3 "Time of the accident"

	3.1	3.2	Estimate
3.1	1	2	0,667
3.2	1/2	1	0,333

Table 4.4 Pair-wise comparison matrix for parameter 4 "Weather conditions"

	4.1	4.2	4.3	4.4	4.5	Estimate
4.1	1	3	2	2	4	0,326
4.2	1/3	1	1/3	1/3	2	0,109
4.3	1/2	3	1	1/2	3	0,217
4.4	1/2	3	2	1	4	0,285
4.5	1/4	1/2	1/3	1/4	1	0,063

Table 4.5 Pair-wise comparison matrix for parameter 5 "Driver's condition"

	5.1	5.2	Estimate
5.1	1	1/3	0,25
5.2	3	1	0,75

Table 4.6 Pair-wise comparison matrix for parameter 6 "Trigger of the accident"

	6.1	6.2	6.3	6.4	6.5	6.6	6.7	Estimate
6.1	1	1	2	3	5	3	6	0,253
6.2	1	1	2	3	5	3	6	0,253
6.3	1/2	1/2	1	2	4	2	5	0,181
6.4	1/3	1/3	1/2	1	3	1/2	3	0,105
6.5	1/5	1/5	1/4	1/3	1	1/3	2	0,052
6.6	1/3	1/3	1/2	2	3	1	3	0,123
6.7	1/6	1/6	1/5	1/3	1/2	1/3	1	0,033

Table 4.7 Pair-wise comparison matrix for parameter 7 "Cause of the accident"

	7.1	7.2	7.3	7.4	7.5	Estimate
7.1	1	3	1/5	1/4	3	0,144
7.2	1/3	1	1/6	1/5	2	0,072
7.3	5	6	1	4	7	0,446
7.4	4	5	1/4	1	5	0,296
7.5	1/3	1/2	1/7	1/5	1	0,042

Table 4.8 Independent probabilities of parameter alternatives

Independent probability estimates						
Type	Place	Time	Weather	Driver's condition	Trigger	Cause
1	2	3	4	5	6	7
0,273	0,311	0,667	0,326	0,25	0,253	0,144
0,484	0,159	0,333	0,109	0,75	0,253	0,072
0,177	0,282		0,217		0,181	0,446
0,066	0,078		0,285		0,105	0,296
	0,036		0,063		0,052	0,042
	0,078				0,123	
	0,056				0,033	

Solving the system of Eq. (3.8) provides the morphological table with calculated final probability values (Table 4.10).

The methods of countering traffic accidents are divided into two groups (Table 4.11).

(A) prevention methods: video surveillance with photo-capturing of violations; with traffic violation photo-capturing; speed traps; traffic lane separators; speed bumps; popularization of traffic rules abidance;
(B) response methods: arrangement of special road worker crews; manual traffic lights management; informing of other drivers about the place of accident.

The dependency matrix conveying the efficiency of the methods for dealing with traffic accidents with specific characteristic parameters is built (Table 4.12).

The probability estimates of the traffic accident parameter alternatives (Table 4.10) and the consistency matrix (Table 4.12) are used to calculate the efficiencies of each of the methods for dealing with the traffic accidents (Table 4.13, Fig. 4.1).

Among the considered methods, the most efficient regarding the whole multitude of possible accidents are video surveillance with photo-capturing of violations; with traffic violation photo-capturing and speed traps as prevention methods, and arrangement of special road worker crews as response methods. It is important to note that MMAM considers only the distribution of different types of accidents, not the overall number of accidents. Universal methods such as reducing traffic intensity by constructing car tunnels and underground parking lots, are still the best method of decreasing the total quantity of accidents.

Table 4.9 The cross-consistency matrix

	1.1	1.2	1.3	1.4	2.1	2.2	2.3	2.4	2.5	2.6	2.7	3.1	3.2	4.1	4.2	4.3	4.4	4.5	5.1	5.2	6.1	6.2	6.3	6.4	6.5	6.6	6.7
2.1		0,3	0,5																								
2.2																											
2.3		0,3																									
2.4																											
2.5																											
2.6			−0,3																								
2.7			0,3																								
3.1																											
3.2											0,1																
4.1																											
4.2		0,1																									
4.3	0,3	0,3																									
4.4	0,7	0,7	0,5																								
4.5	0,5	0,3																									
5.1																											
5.2																											
6.1	0,5	0,5	0,5		0,5	0,5	0,5			−0,5	0,3				0,3	0,3	0,5	0,3									
6.2	0,3	0,7	0,3			0,3	0,3	0,5	0,5		−0,7				0,1	0,1		0,3									
6.3		0,5	0,5		0,5	−1	−1	−1	−1	−1	−1																
6.4		0,5			0,5	0,3	0,5	0,5	0,5	0,3							0,1										

(continued)

Table 4.9 (continued)

	1.1	1.2	1.3	1.4	2.1	2.2	2.3	2.4	2.5	2.6	2.7	3.1	3.2	4.1	4.2	4.3	4.4	4.5	5.1	5.2	6.1	6.2	6.3	6.4	6.5	6.6	6.7
6.5		0,3			0,3			0,3	0,3		−0,3				0,1	0,1		0,1									
6.6	0,5		0,3								0,3																
6.7																											
7.1	0,3	0,3	0,3										0,3		0,3	0,3	0,3	0,3		0,3	−0,1						
7.2	0,3	0,3	0,3													0,3	0,5			0,3		0,3			−0,5		
7.3	0,3	0,3	0,3										0,3							0,5	0,5	0,5	0,5	0,5	0,3	0,5	−1
7.4				0,7																				0,3	0,7		−1
7.5																											0,7

Table 4.10 Calculated final probabilities of parameter alternatives

Final probability estimates						
Type	Place	Time	Weather	Driver's condition	Trigger	Cause
1	2	3	4	5	6	7
0,273	0,331	0,659	0,296	0,246	0,278	0,158
0,478	0,156	0,341	0,106	0,754	0,256	0,085
0,188	0,284		0,218		0,152	0,46
0,06	0,076		0,31		0,121	0,261
	0,036		0,07		0,055	0,037
	0,066				0,117	
	0,051				0,02	

Table 4.11 Traffic accident countering methods

Prevention methods	Response methods
A	B
(A.1) Video surveillance with photo-capturing of violations	(B.1) Arrangement of special road worker crews
(A.2) Speed traps	(B.2) Manual traffic lights management
(A.3) Traffic lane separators	(B.3) Informing of other drivers about the place of accident
(A.4) Speed bumps	
(A.5) Popularization of traffic rules abidance	

4.2 Traffic Jam Scenarios Evaluation

Another traffic problem that was studied with the MMAM was the problem of traffic jams, which is increasingly urgent for large cities and metropolises. Two-stage MMAM was conducted for analyzing the methods of leveraging this issue. The following parameters that characterize traffic jams were chosen:

1. time of a traffic jam appearance: peak hours; non-peak hours;
2. city region: city center; main passageways to the city center; regions far from city center;
3. place of a traffic jam: traffic lanes and bridges; regulated crossings; unregulated crossings; road junctions; complex jams over several traffic nodes;
4. regularity: regular (occur often in the same place, likely each day); irregular (occur spontaneously, without consistent patterns);
5. cause of a traffic jam: low throughput; traffic accident; vehicle-related—violations, malfunctioning; weather; road works; special events (festivities, rallies, etc.); parked vehicles; other causes.

The parameters are gathered in a morphological table (Table 4.14).

Table 4.12 The dependency matrix for methods of dealing with traffic accidents

	A.1	A.2	A.3	A.4	A.5	B.1	B.2	B.3
1.1			−0,1			0,5	0,1	0,5
1.2						0,7	0,5	0,7
1.3						0,3		0,3
1.4				−0,3		0,5		
2.1	0,3	0,3		0,5	0,3	0,5	0,9	0,7
2.2	0,1	0,3	0,3	0,3	0,1	0,3	−1	0,5
2.3	0,3	0,5	0,5	0,3	0,1	0,5	−1	0,5
2.4	0,3	0,3	0,5	0,3		0,5	−1	0,7
2.5	0,3	0,3	0,5	0,3		0,5	−1	0,7
2.6	0,3		–0,5	0,5		0,3	−1	0,3
2.7	–0,7	–0,7	–0,7	0,7			−1	−0,1
3.1								
3.2	–0,3							
4.1	0,5	0,3						
4.2	–0,1							
4.3	–0,1							
4.4	0,5							
4.5	–0,5	–0,3						
5.1								
5.2					0,5			
6.1	0,5	0,9		0,5	0,1			
6.2	0,5	−1	0,7	−0,5	0,3			
6.3	0,5	−1	−1	−1	0,3			
6.4	0,5	−1	0,5	−0,5	0,3			
6.5	0,5		−1	−1	0,3			
6.6	0,5	−1	0,5	0,5	0,3			
6.7	−1	−1	−1	−1	−0,7			
7.1								
7.2				−0,3				
7.3					0,5			
7.4								
7.5								

Table 4.13 The calculated efficiencies of methods for dealing with traffic accidents

Methods for preventing traffic accidents		Methods for reduction of accidents' consequences	
Method	Efficiency	Method	Efficiency
(A.1) Video surveillance with photo-capturing of violations	0,339	(B.1) Arrangement of special road worker crews	0,529
(A.2) Speed traps	0,228	(B.2) Manual traffic lights management	0,114
(A.3) Traffic lane separators	0,134	B.3) Informing other drivers about the place of accident	0,357
(A.4) Speed bumps	0,131		
(A.5) Popularization of traffic rules abidance	0,169		

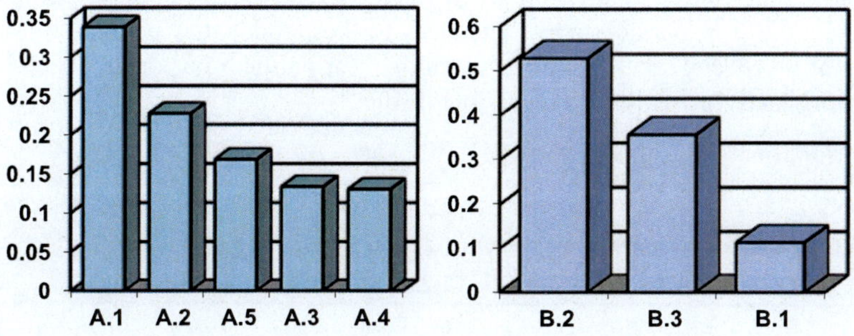

Fig. 4.1 Ranked expected efficiencies for parameters A, B

Table 4.14 Morphological table for traffic jams

Characteristic parameters				
1. Time	2. Region	3. Place	4. Regularity	5. Cause
1.1. Peak hours	2.1. City center	3.1. Traffic lanes and bridges	4.1. Regular	5.1. Low throughput
1.2. Non-peak hours	2.2. Passageways to city center	3.2. Regulated crossings	4.2. Irregular	5.2. Traffic accident
	2.3. Far from city center	3.3. Unregulated crossings		5.3. Violations, malfunctioning
		3.4. Road junctions		5.4. Weather
		3.5. Several nodes		5.5. Road works
				5.6. Special events
				5.7. Parked vehicles
				5.8. Other

The experts estimated independent probabilities of alternatives using pair-wise comparison. The results are presented in Tables 4.15, 4.16, 4.17, 4.18 and 4.19.

The pair-wise comparisons are processed to obtain the preliminary probability estimates (Table 4.20).

The cross-consistency matrix is shown in Table 4.21. Solving the system of Eq. (3.8) provides the morphological table with calculated final probability values (Table 4.22).

The calculated distributions of probabilities were used to analyze and compare the efficiency of traffic jam prevention measures. Three groups of measures were picked out: construction measures, informational measures, and organizational measures.

Considered construction measures for countering traffic jams included:

- road widening—creating new traffic lanes to increase throughput;
- constructing new tunnels, junctions, and bridges;
- constructing transit tunnels, highways, and freeways without crossings that connect city's important nodes. The entry to these roads may be paid;
- separating special lanes, "express corridors" for public transport;
- creating lanes or roads with reversible traffic.

Informational measures included:

Table 4.15 Pair-wise comparison matrix for parameter 1 "Type of the accident"

	1.1	1.2	Estimate
1.1	1	6	0,857
1.2	1/6	1	0,143

Table 4.16 Pair-wise comparison matrix for parameter 2 "City region"

	2.1	2.2	2.3	Estimate
2.1	1	2	4	0,535
2.2	1/2	1	3	0,344
2.3	1/4	1/3	1	0,121

Table 4.17 Pair-wise comparison matrix for parameter 3 "Place of a traffic jam"

	3.1	3.2	3.3	3.4	3.5	Estimate
3.1	1	1/5	1/6	1/6	1/7	0,036
3.2	5	1	1/3	1/2	1/3	0,155
3.3	6	3	1	2	1/2	0,27
3.4	6	2	1/2	1	1/2	0,216
3.5	7	3	2	2	1	0,324

Table 4.18 Pair-wise comparison matrix for parameter 4 "Regularity"

	4.1	4.2	Estimate
4.1	1	3	0,75
4.2	1/3	1	0,25

Table 4.19 Pair-wise comparison matrix for parameter 5 "Cause of a traffic jam"

	5.1	5.2	5.3	5.4	5.5	5.6	5.7	5.8	Estimate
5.1	1	3	4	7	5	9	4	9	0,266
5.2	1/3	1	2	7	4	8	2	8	0,205
5.3	1/4	1/2	1	6	4	8	3	8	0,195
5.4	1/7	1/7	1/6	1	1/3	3	1/4	5	0,064
5.5	1/5	1/4	1/4	3	1	5	1/2	6	0,103
5.6	1/9	1/8	1/8	1/3	1/5	1	1/6	2	0,026
5.7	1/4	1/2	1/3	4	2	6	1	6	0,127
5.8	1/9	1/8	1/8	1/5	1/6	1/2	1/6	1	0,015

Table 4.20 Independent probabilities of parameter alternatives

Characteristic parameters				
1. Time	2. Region	3. Place	4. Regularity	5. Cause
0,857	0,535	0,036	0,75	0,266
0,143	0,344	0,155	0,25	0,205
	0,121	0,27		0,195
		0,216		0,064
		0,324		0,103
				0,026
				0,127
				0,015

- systems of centralized traffic lights control as a part of automated system of city's traffic management;
- intellectual traffic lights that adapt to the information about movement intensity from special detectors;
- electronic boards that inform traffic participants about jams so that they can alter their routes;
- special radio channel that informs about traffic jams;
- interaction with navigational systems for displaying traffic intensity on electronic maps.

Organizational measures included:

Table 4.21 The cross-consistency matrix

	1.1	1.2	2.1	2.2	2.3	3.1	3.2	3.3	3.4	3.5	4.1	4.2
2.1	0,3	0,3										
2.2	0,7											
2.3												
3.1			−0,7	0,3								
3.2			0,7	0,3								
3.3			−0,3									
3.4			0,7	0,3								
3.5	0,3	−0,3	0,7	0,3	−0,3							
4.1	0,7		0,7	0,7			0,3	0,3	0,3	0,3		
4.2				0,3								
5.1	0,7	0,3	0,7	0,3		0,3				0,3	1	
5.2								0,3			−1	0,7
5.3			0,3				0,3	0,7	0,3		−1	0,7
5.4										0,3	−1	0,3
5.5			0,3			0,3				0,3	−0,7	0,7
5.6	−0,3		0,3	−0,3	−0,7						−1	
5.7			0,7	−0,3		0,3			0,3	0,3	0,7	0,3
5.8												

- reducing traffic intensity by limiting allowance for cars (for example, cars with odd or even last number of car plate are allowed on odd or even days of a month, respectively—similar rules are established in some cities—Athens, Mexico City, etc.);
- paid entry to city center (e.g., London, Rome);
- increased fines on violations;
- encouraging car pools, public transport, "no car" days, bicycles, etc.;
- forbidding left turn on unregulated crossings;
- limiting commercial transportation to night time;
- varying schedules for companies to avoid peak hours.

The measures for preventing traffic jams are shown in the strategy MT (Table 4.23).

The expert-assessed dependency matrix for parameters F_6, F_7, F_8 on the parameters of scenario MT is shown in Table 4.24.

Using the previously calculated probability distributions for traffic jams, and the dependency matrix, the expected efficiencies of prevention measures are calculated (Table 4.25).

As a result of the conducted research, a conclusion can be made that according to Table 4.22, low throughput and parked vehicles constitute the most significant cause of traffic jam occurrence. The most potentially efficient prevention measures are: among construction measures—constructing new tunnels, junctions, bridges, and

Table 4.22 Calculated final probability values for traffic jam parameters

Characteristic parameters									
1. Time		2. Region		3. Place		4. Regularity		5. Cause	
1.1. Peak hours	0.793	2.1. City center	0.480	3.1. Traffic lanes and bridges	0.065	4.1. Regular	0.515	5.1. Low throughput	0.379
1.2. Non-peak hours	0.207	2.2. Passageways to city center	0.474	3.2. Regulated crossings	0.233	4.2. Irregular	0.485	5.2. Traffic accident	0.092
		2.3. Far from city center	0.045	3.3. Unregulated crossings	0.127			5.3. Violations, malfunctioning	0.100
				3.4. Road junctions	0.265			5.4. Weather	0.033
				3.5. Several nodes	0.310			5.5. Road works	0.095
								5.6. Special events	0.002
								5.7. Parked vehicles	0.293
								5.8. Other	0.006

Table 4.23 Traffic jam countering methods

6. Construction measures	7. Informational measures	8. Organizational measures
6.1. Road widening	7.1. Centralized traffic lights control	8.1. Limiting allowance for cars
6.2. New tunnels, junctions, bridges	7.2. Intellectual traffic lights	8.2. Paid entry to city center
6.3. Transit tunnels, highways, freeways	7.3. Electronic boards	8.3. Increased fines on violations
6.4. "Express corridors" for public transport	7.4. Special radio channel	8.4. Encouraging car pools, public transport, "no car" days, bicycles
6.5. Reversible traffic roads	7.5. Interaction with navigational systems	8.5. Forbidding left turn
		8.6. Commercial transportation at night time
		8.7. Varying schedule for companies

creating roads with reversible traffic; among the informational measures—organizing centralized traffic lights control and installing intellectual traffic lights so that the traffic jam can be efficiently managed in automated mode; among the organizational measures—the most radical methods were assessed as the most efficient—limiting car allowance and paid entry to the city center. But it should be noted that the efficiency was evaluated according to the measures' capacity of dealing with traffic jams, not their feasibility.

4.3 Predictive Assessment of Geological Environment Favorability for Urban Underground Construction

Continuous growth of large cities is a display of consistent historical patterns and leads not only to the increasing size of metropolises but also to significant complication of their functional and spatial organization as well. In many cases, the capacity of expanding "upwise and broadwise" is nearly exhausted. Solving a set of acute problems related to the intensive growth of metropolises, including territorial, transport, power supply, ecological, and other problems, can be achieved by developing the urban underground space. The concept of sustainable growth for large cities holds a special place for underground development, as the underground infrastructure increases the quality of life and ecological safety much more than a similar structure on surface (Vähäaho 2014; Gilbert et al. 2013). Despite the systemic advantages and significant prospects of developing underground space, the pace of urban underground construction does not satisfy the needs of modern metropolises. One

Table 4.24 Dependency matrix for traffic jam prevention measures

	6.1	6.2	6.3	6.4	6.5	7.1	7.2	7.3	7.4	7.5	8.1	8.2	8.3	8.4	8.5	8.6	8.7
1.1			0,3		0,7									0,3			0,3
1.2																	−0,7
2.1	−0,7		−0,7	−0,3	−0,7	0,3	0,3				0,3	0,7				−0,3	0,3
2.2		0,3	0,3		0,7	0,3	0,3				0,3	0,3			−0,3	0,3	0,3
2.3		0,3					0,3	−0,3			0,3					0,3	
3.1	0,3	0,3			0,3												
3.2	−0,3	0,3				0,7	0,7										
3.3	−0,3	0,3				−1	−1								0,7		
3.4						0,3	0,3								0,7		
3.5	0,3	0,3		0,3	0,3	0,7		0,3			0,3	0,3		0,3			
4.1			0,3			0,3	0,3				0,3			0,3			0,3
4.2							0,3		−0,3								
5.1	0,7	0,3	0,3	−0,3	0,3						0,7	0,7		0,3		0,3	0,3
5.2					−0,3		0,3	0,3	0,3	0,3	0,3				0,3		
5.3					−0,3		0,3				0,3		0,7		−0,3		
5.4																	
5.5						0,3		0,3	0,3	0,3							
5.6								0,3	0,3	0,3							
5.7	0,3										0,7		0,7				
5.8																	

Table 4.25 Expected efficiencies for traffic jam prevention methods

6. Construction measures		7. Informational measures		8. Organizational measures	
6.1. Road widening	0.149	7.1. Centralized traffic lights control	0.240	8.1. Limiting allowance for cars	0.180
6.2. New tunnels, junctions, bridges	0.291	7.2. Intellectual traffic lights	0.238	8.2. Paid entry to city center	0.170
6.3. Transit tunnels, highways, freeways	0.183	7.3. Electronic boards	0.196	8.3. Increased fines on violations	0.127
6.4. "Express corridors" for public transport	0.136	7.4. Special radio channel	0.144	8.4. Encouraging car pools, public transport, "no car" days, bicycles	0.157
6.5. Reversible traffic roads	0.241	7.5. Interaction with navigational systems	0.182	8.5. Forbidding left turn	0.110
				8.6. Commercial transportation at night time	0.108
				8.7. Varying schedule for companies	0.147

of the main reasons for this situation is that developing underground space holds considerable risks caused by insufficient information about the state and properties of geological environment at the stage of underground infrastructure planning and making decisions regarding the advisability of investment under the conditions of incomplete information. Thus, preliminary estimation of favorability of urban territories for underground development is an urgent issue for increasing volumes of underground construction in cities.

An element of urban territory suitable for morphological evaluation is a (potential) construction site. Forming the morphological table of a construction site requires its classification by different characteristics relevant to the decision. Each slice of classification becomes a parameter of the MT, and its ranges or values become the alternatives of this MT parameter. The actual number of parameters in this problem is actually very high for convenient analysis, which is why some sets of characteristics that have a similar impact on the decision, were aggregated into single parameters. The final presentation of the MT for the problem is given in Table 4.26.

Let us briefly review the selected parameters, which set the goal of maximally objective representation of geological environment's and technogenic factors' impact on the risks of underground construction and maintenance of underground objects (Haiko 2015). Parameters 1–4 and 6 specify the threats to stability (reliability) of an

Table 4.26 Morphological table for a construction site

Parameter	Parameter alternative
1. Level of dynamic load	1.1. Low (46–53 dB)
	1.2. Medium (53–73 dB)
	1.3. Increased (73–96 dB)
	1.4. High (over 96 dB)
2. Static load from surface buildings	2.1. Insignificant ($K_{sl} < 1$)
	2.2. Medium ($1 < K_{sl} < 2$)
	2.3. Increased ($2 < K_{sl} < 3{,}5$)
	2.4. High ($K_{sl} > 3{,}5$)
3. Static load from soil	3.1. Insignificant ($K_{mas} < 0{,}05$, MPa)
	3.2. Medium ($0{,}05 < K_{mas} < 0{,}3$, MPa)
	3.3. High ($0{,}3 < K_{mas} < 0{,}5$, MPa)
	3.4. Very high ($K_{mas} > 5$, MPa)
4. Influence of existing underground objects	4.1. Absent (distance over 50 m)
	4.2. Slight (distance 20–50 m)
	4.3 Significant (distance 10–20 m)
	4.4 Hazardous (distance less than 10 m)
5. Genetic type and lithologic composure of soil	5.1. Unweathered clays and average density sands
	5.2. Technogenic deposits (alluvial and bulk types)
	5.3. Deluvial clay soils (water-saturated), water-saturated overfloodplain sands
	5.4. Sedentary soils, soils with special properties (loess, peat, silt)
6. Effective soil strength	6.1. Very strong soils > 300 kPa
	6.2. Strong soils 200–300 kPa
	6.3. Average strength soils 150–200 kPa
	6.4. Relatively strong soils <150 kPa
7. Influence of aquifers and perched groundwater	7.1. Water-bearing horizons at $P\text{-}N_{1np}$
	7.2. Groundwater depth > 3 m, pressurized groundwater >10 m
	7.3. Groundwater depth < 3 m, pressurized groundwater <10 m
	7.4. Flooded areas with groundwater level up to 1 m present
8. Landscape type and morphometrics	8.1. Flat areas of overfloodplain terraces, morainic-glacial plains
	8.2. Slightly tilted overfloodplain terraces, watershed ares

(continued)

Table 4.26 (continued)

Parameter	Parameter alternative
	8.3. Small river valleys, slightly irregular slopes, high floodplain
	8.4. Slope areas with ravines and steep banks, low floodplain
9. Geological engineering processes	9.1. Absent
	9.2. Stabilized
	9.3. Low displacement processes
	9.4. Active manifestations of subsidence, underflooding, gravitational processes
10. Geotechnologies of underground construction	10.1. Open
	10.2. Underground

underground structure, and reflect its subjection to dynamic loads (roads, railway, subway, industrial facilities, etc.), static load from surface buildings (taking into account the density of urban structures with a coefficient related to the number of stories in buildings); load from soil pressure on the structure's lining (taking into account soil density, depth of soil pillar, humidity and class coefficients of soils); geomechanical disturbance of soil array from other underground structures and resistance of soil to loads. Parameters 5, 7–9 describe the processes and properties of geological environment that significantly influence the cost and speed of construction, the conditions for repair-free maintenance of underground structures, and ecological safety. Authors acknowledge the possible interrelations between the groups of geomechanical and geological factors; however, both groups are strictly necessary for analysis and will be equally applied to all studied territories (construction sites). The accepted factor ranges correspond to the four levels of geological environment's favorability to mastering underground space: very favorable, favorable, insufficiently favorable, and unfavorable.

The second MT contained decision parameters for a described construction site. A total of six parameters are chosen (Table 4.27).

General characteristics include parameters A, B, and C. Parameter A (Site suitability) is integral, with the alternative A.1 (Suitable) encompassing very favorable, favorable, and insufficiently favorable geological environments, and the alternative A.2 (Not suitable)—unfavorable environment that has high risk indicators (Parameters D–F). Parameter B (Object scale) references the types of designed underground structures in the city infrastructure, in particular: sewer and power communications (alternative B.1), traffic passageways (alternative B.2), multifunctional underground chamber structures (alternative B.3), and chambers with large cross-sections: underground malls, sport objects, power supply stations, industrial facilities, etc. (alternative B.4). As various geological environments may favor different cross-sections, decision-making regarding the construction site should factor the alternatives of Parameter B (Object scale). Recommended construction depth (Parameter C) is

Table 4.27 Decision alternatives regarding a construction site

General analysis

A. Site suitability	B. Object scale	C. Construction depth
A.1. Suitable	B.1. Cross-section up to 10 m^2	C.1. 0–10 m
A.2. Not suitable	B.2. Cross-section up to 35 m^2	C.2. 10–20 m
	B.3. Cross-section up to 70 m^2	C.3. 20–50 m
	B.4. Cross-section up to and over 70 m^2	C.4. beneath 50 m

Construction risks

D. Risk factor	E. Risk degree	F. Risk level
D.1. Construction failure, malfunction	E.1. <3%	F.1. 0,1–5% Q
D.2. Dangerous influence on surface or neighboring underground objects	E.2. 3–10%	F.2. 5–20% Q
D.3. Initiating displacements	E.3. 10–20%	F.3. 20–50%Q
D.4. Underflooding	E.4. 20–50%	F.4. >50% Q
D.5. Ecological risks	E.5. >50%	
D.6. Transport problems		
D.7. Increasing construction and operation cost		

related to the functional purpose of the structure and chosen geoengineering technology, it influences the forming of loads on the structure's lining from soil pressure, and static and dynamic impacts.

The Parameters D, E, and F refer to the risks of underground construction. Risk factor alternatives include construction failure, malfunction (alternative D.1), dangerous influence on surface or neighboring underground objects (alternative D.2), initiating displacements (alternative D.3), underflooding (alternative D.4), ecological risks (alternative D.5), transport problems (alternative D.6), and increasing construction and operation cost (alternative D.7). Risk degree (Parameter E) shows the probability of emergence of an undesired event (i.e., alternatives D.1–D.7), and risk level (Parameter F) estimates the economic losses in case of undesired events as a fraction of initial structure's cost Q.

It should be noted that not all of the MT parameters directly influence each other. At the CCM filling stage, the questions for pairs of parameters that obviously do not have relation were excluded to lessen the load on experts. Pairs of parameters from Table 4.26, the dependencies between which were actually estimated, are shown in Table 4.28.

The pairs of parameters, where the relation was established, had the dependences between their alternatives assessed using the CCM filling procedures described in previous chapters. The same was accomplished to fill the dependence matrix that connects the first and second MTs.

Table 4.28 Matrix of interconnections between parameters of the construction site MT ("+" denotes direct influence)

Parameter	1	2	3	4	5	6	7	8	9	10
1			+	+					+	
2			+			+	+		+	+
3	+	+		+	+	+	+	+	+	+
4	+		+				+		+	+
5			+			+	+	+	+	
6		+	+		+					
7		+	+	+	+				+	
8			+		+				+	+
9	+	+	+	+	+		+	+		
10		+	+	+				+		

The created model allows to evaluate construction sites. The input data for the model comprises of the expert opinions regarding the initial estimates of the first MT (Table 4.26), i.e., the site characteristics. The evaluation procedure consists of the following main steps:

1. obtaining information from experts using the questionnaire;
2. translating the experts' answers into numerical form and calculating probabilities for the first MT taking into account the dependences between its parameters;
3. calculating weights of the second MT on the base of the multitude of configurations from the first MT, and the dependence matrix.

At the first stage, the expert gets a questionnaire regarding a studied site. This questionnaire contains questions regarding the appropriateness of an alternative for the studied site. A possible form of these questions is as follows:

1. Please specify how correct are the different values of the parameter ***"Level of dynamic load"*** *for the studied site*

1.1. Is it true that the considered construction site has ***Level of dynamic load—Low (46–53 dB)***

Unequivocally untrue	Mostly untrue	Slightly untrue	Can be true or untrue	Slightly true	Mostly true	Unequivocally true

1.2. Is it true that the considered construction site has ***Level of dynamic load—Average (53–73 dB)***

Unequivocally untrue	Mostly untrue	Slightly untrue	Can be true or untrue	Slightly true	Mostly true	Unequivocally true

1.3. Is it true that the considered construction site has ***Level of dynamic load—Increased (73–96 dB)***

Unequivocally untrue	Mostly untrue	Slightly untrue	Can be true or untrue	Slightly true	Mostly true	Unequivocally true

1.4. Is it true that the considered construction site has ***Level of dynamic load—High (> 96 dB)***

Unequivocally untrue	Mostly untrue	Slightly untrue	Can be true or untrue	Slightly true	Mostly true	Unequivocally true

An expert selects the answers that, to his opinion, characterize each alternative most accurately for the studied site. If precise information regarding a site is present, the expert may choose "Unequivocally true" for one of the alternatives, and "Unequivocally untrue" for the others. Such cases constitute the fixed parameter problem for MMAM, which was covered in Sect. 3.4. The number of questions corresponds to the total number of alternatives in the first MT. For the task of evaluating a construction site, the questionnaire consists of 38 questions.

To test the functioning of the developed model, two underground parking lot sites in Kyiv with different characteristics were taken.

The first construction site is found in the Shevchenkivsky district at the Peremohy avenue. From a geomorphologic viewpoint the site is placed in the bounds of two geomorphologic elements—first overfloodplain terrace of Lybid river and its valley slope. Geological composure of the studied site down to the depth of 50 m comprises of alluvial and fluvioglacial deposits including silty sandy loam, sand, peat, low- and medium-clay loam that lie upon bedrock of the Kyiv Paleocene suite represented by marly loam and marly clay. In turn, Kyiv suite soils lie upon sandy loam and sand of the Paleocene Buchak suite. On the surface, alluvial and Paleocene deposits are covered by bulk ground. The studied soil characteristics, obtained from 11 boreholes, were used in the morphological model construction.

The second construction site is also found in the Shevchenkivsky district between the Bulvarno-Kudriavska and Honchara streets. The studied territory is placed on the left slope of the Lybid River valley. The primary landscape surface is significantly heightened by bulk ground up to Bulvarno-Kudriavska street marks at this place, additionally, the embankment is fenced off by a supporting rubble stone wall. Geological composure of the studied site down to the investigated depth of 36 m comprises of: on the territory surface—modern bulk and Holocene deluvial deposits with underlying Holocene and upper Pleistocene deluvial-shearing deposits that occasionally cover deluvial soils of early Pleistocene. Under a substational stratum of deluvial and bulk soils, a partially washed-out stratum of the Poltava series of lower Neocene is found. The investigated slope is comprised of bulk soils with poor filtration properties, which poses a probability of forming individual water lenses in sandy layers of bulk, or even forming perched groundwater spots in upper layers of cross-section in case the natural draining of the slope is interrupted. The studied soil characteristics and technogenic influence were used in the morphological model construction.

The input data from experts and the calculated in the MMAM procedure estimates that take into account the interrelation between parameters, are given in Table 4.29. The normalized expert estimate (probability) of an alternative $a_j^{(i)}$ is shown in $p_j^{\prime(i)}$ column; the estimate that was calculated using the cross-consistency matrix is shown in the column $p_j^{(i)}$. The difference between the values demonstrates the influence of other parameters on the probability of choosing a respective alternative.

The calculated weights of parameter alternatives for the second MT are shown in Table 4.30.

The results can be presented in a more intuitive form of charts or graphs (Figs. 4.2 and 4.3).

Pie charts (Fig. 4.2) demonstrate the most likely risk factors for the construction sites. In both cases the biggest hazard lies in initiating displacements (0,241 and 0,324 respectively), caused by the Lybid River influence and a slope relief for the second site which is prone to landslides. The risk factor of increasing construction and operation cost has the second biggest value for site 1 (0,202), corresponding to the more difficult geomechanical situation compared to site 2, where the ecological risks have a bigger impact (0,3). Both sites also have substantial risks of territory underflooding (0,154 and 0,214, respectively). Other risk factors are less relevant.

The defining factors for risk estimation are the parameters E (risk degree, i.e., the probability of the risk situation) and F (risk level, i.e., the expected financial and economic loss in case the risk situation happens—for example, the cost of repair). The comparison graphs for risk degree (Fig. 4.3, left) point at the largest likelihood of unfavorable scenarios at 3–10% (with weights 0,502 and 0,625, respectively). Additionally, the likelihood of high risks (20–50%) is less than 0,072 for site 1 and nearly equals zero for site 2, assuming that the conditions are largely favorable for construction. The assessment of possible financial losses in case of unfavorable scenarios (although they have low enough likelihood), can be performed by studying risk level graphs (Fig. 4.3, right). The most probable situation is that the financial risks have the 5–20% level of construction cost, which is less than the average of the estimation scale. Thus, both sites are favorable for underground construction, which is confirmed by absolute weights of parameter A in Table 4.30, with "Favorable" alternative having values of 0,688 and 0,993, respectively.

Table 4.29 Input data and calculated estimates for parameter alternatives of the first MT

Parameter	Alternative	Estimates			
		Site 1		Site 2	
		$p_j'^{(i)}$	$p_j^{(i)}$	$p_j'^{(i)}$	$p_j^{(i)}$
1. Level of dynamic load	1.1. Low (46–53 dB)	0,000	0,000	0,593	0,707
	1.2. Medium (53–73 dB)	0,194	0,388	0,259	0,263
	1.3. Increased (73–96 dB)	0,361	0,286	0,148	0,029
	1.4. High (over 96 dB)	0,444	0,326	0,000	0,000
2. Static load from surface buildings	2.1. Insignificant ($К_{СН}$ < 1)	0,000	0,000	0,000	0,000
	2.2. Medium (1 < $К_{СН}$ < 2)	0,000	0,000	1,000	1,000
	2.3. Increased (2 < $К_{СН}$ < 3,5)	0,000	0,000	0,000	0,000
	2.4. High ($К_{СН}$ > 3,5)	1,000	1,000	0,000	0,000
3. Static load from soil	3.1. Insignificant ($К_{mas}$ < 0,05, MPa)	0,351	0,030	0,432	0,324
	3.2. Medium (0,05 < $К_{mas}$ < 0,3, MPa)	0,351	0,717	0,351	0,639
	3.3. High (0,3 < $К_{mas}$ < 0,5, MPa)	0,189	0,189	0,108	0,025
	3.4. Very high ($К_{mas}$ > 5, MPa)	0,108	0,063	0,108	0,012
4. Influence of existing underground objects	4.1. Absent (distance over 50 m)	0,000	0,000	0,800	0,714
	4.2. Slight (distance 20–50 m)	0,800	0,479	0,200	0,286
	4.3 Significant (distance 10–20 m)	0,200	0,521	0,000	0,000
	4.4 Hazardous (distance less than 10 m)	0,000	0,000	0,000	0,000
5. Genetic type and lithologic composure of soil	5.1. Unweathered clays and average density sands	0,108	0,107	0,108	0,170
	5.2. Technogenic deposits (alluvial and bulk types)	0,351	0,372	0,351	0,450
	5.3. Deluvial clay soils, overfloodplain sands	0,432	0,434	0,432	0,342
	5.4. Sedentary soils, soils with special properties	0,108	0,087	0,108	0,038
6. Effective soil strength	6.1. Very strong soils > 300 kPa	0,000	0,000	0,000	0,000
	6.2. Strong soils 200–300 kPa	0,000	0,000	0,333	0,209
	6.3. Average strength soils 150–200 kPa	0,200	0,352	0,533	0,670

(continued)

Table 4.29 (continued)

Parameter	Alternative	Estimates			
		Site 1		Site 2	
		$p_j'^{(i)}$	$p_j^{(i)}$	$p_j'^{(i)}$	$p_j^{(i)}$
	6.4. Relatively strong soils <150 kPa	0,800	0,648	0,133	0,120
7. Influence of aquifers and perched groundwater	7.1. Water-bearing horizons at P-N_{1np}	0,000	0,000	0,143	0,098
	7.2. G/water depth > 3 m, pressurized g/water > 10 m	0,121	0,384	0,571	0,672
	7.3. G/water depth < 3 m, pressurized g/water < 10 m	0,485	0,552	0,143	0,126
	7.4. Flooded areas with g/water level up to 1 m present	0,394	0,065	0,143	0,104
8. Landscape type and morphometrics	8.1. Flat overfloodplain terraces, morainic-glacial plains	0,571	0,476	0,000	0,000
	8.2. Slightly tilted overfloodplain terraces	0,143	0,177	0,121	0,382
	8.3. Small river valleys, high floodplain	0,143	0,275	0,394	0,476
	8.4. Slope areas with ravines and steep banks	0,143	0,072	0,485	0,142
9. Geological engineering processes	9.1. Absent	0,118	0,018	0,000	0,000
	9.2. Stabilized	0,382	0,346	0,256	0,507
	9.3. Low displacement processes	0,382	0,559	0,410	0,467
	9.4. Active manifestations of subsidence, underflooding, gravitational processes	0,118	0,077	0,333	0,025
10. Geotechnologies	10.1. Open	0,350	0,281	0,448	0,634
	10.2. Underground	0,650	0,719	0,552	0,366

4.4 Model of Assessing Territories for Underground Parking Lots

Among the large diversity of urban underground objects, the underground parking lots stand out especially due to the significant urgency of the problem of parking in central (business) regions of a metropolis, as well as its periphery, where the dwellers of satellite cities and localities commonly switch to the city transport (mainly subway) (Samedov 2011; Haiko 2014; Resin 2013). As shown in the paper (Pankratova 2009), the cluttering of traffic passageways by temporary vehicle parking is an important issue in the goal of eliminating traffic jams and increasing average move speed in

Table 4.30 Alternative parameter weights for the second MT

Parameter	Alternative	Estimate	
		Site 1	Site 2
A. Site suitability	A.1. Suitable	0,688	0,993
	A.2. Not suitable	0,312	0,007
B. Object scale	B.1. Cross-section up to 10 m^2	0,722	0,454
	B.2. Cross-section up to 35 m^2	0,241	0,303
	B.3. Cross-section up to 70 m^2	0,033	0,201
	B.4. Cross-section up to and over 70 m^2	0,005	0,041
C. Construction depth	C.1. 0–10 m	0,053	0,261
	C.2. 10–20 m	0,143	0,307
	C.3. 20–50 m	0,439	0,309
	C.4. beneath 50 m	0,365	0,123
D. Risk factor	D.1. Construction failure, malfunction	0,047	0,002
	D.2. Dangerous influence on surface or neighboring underground objects	0,049	0,006
	D.3. Initiating displacements	0,241	0,324
	D.4. Underflooding	0,154	0,214
	D.5. Ecological risks	0,185	0,300
	D.6. Transport problems	0,122	0,081
	D.7. Increasing construction and operation cost	0,202	0,073
E. Risk degree	E.1. <3%	0,028	0,304
	E.2. 3–10%	0,502	0,625
	E.3. 10–20%	0,382	0,068
	E.4. 20–50%	0,072	0,003
	E.5. >50%	0,017	0,000
F. Risk level	F.1. 0,1–5% Q	0,037	0,562
	F.2. 5–20% Q	0,789	0,422
	F.3. 20–50% Q	0,153	0,015
	F.4. >50% Q	0,020	0,000

large cities. Thus, the role of parking lots becomes even more important. On the other hand, spare sites for constructing open-air parking lots are practically non-existent in the central regions of a metropolis, as former city planning did not foresee the present quantity of cars. Additionally, any "spare" sites were occupied, often using questionable methods, by construction companies for erecting apartment houses, office buildings, or malls. This leads to the conclusion that the parking problem downtown and near terminal subway stations has only one profound solution—constructing underground parking lots. Among the examples of the system approach to this problem one can note nearly simultaneous construction of 41 parking lots in Paris, moreover,

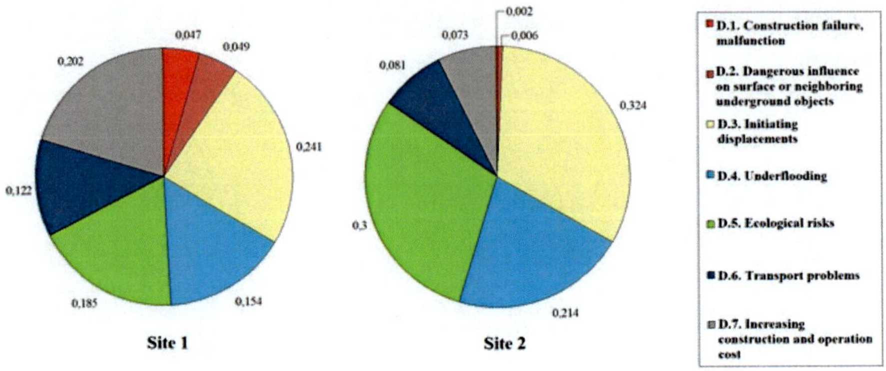

Fig. 4.2 Chart for parameter "D. Risk factor" of the first and second sites

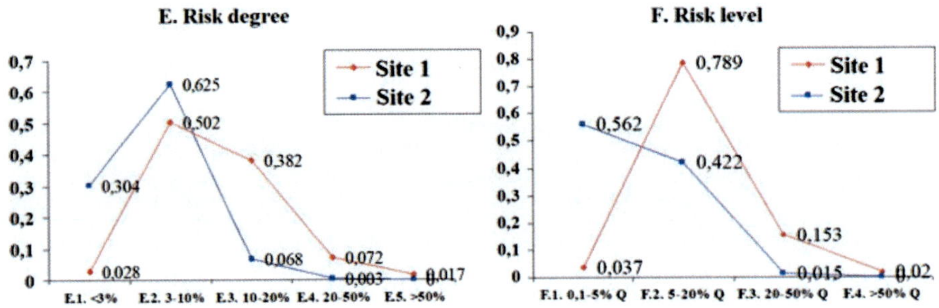

Fig. 4.3 Comparison of weights for alternatives of parameters "E. Risk degree", "F. Risk level" for both sites

the Scientific Coordination Council of Urban Underground Construction, directed by the famous organizer of underground development in European cities E. Utudzhyan, conducted complex and controversial discussions regarding the placement of these underground objects (Kelemen 1985).

The geological environment factors described in previous chapters largely define the cost and risks of constructing and maintaining the structure. However, the problem of reasonable placement of a parking lot is more complex, as it combines both evaluating the influence of geological environment and the structural-functional factors that determine the demand for a parking lot on a set territory and the commercial investment appeal (Ryabkova 2014; Radkevych 2018).

The model for assessing construction sites for parking lots was represented by a network of morphological tables (Fig. 4.4).

The parameters of objects in morphological analysis method were grouped in four separate morphological tables. The first table "I. Site characteristics" contains six parameters with respective alternatives that correspond to the construction site:

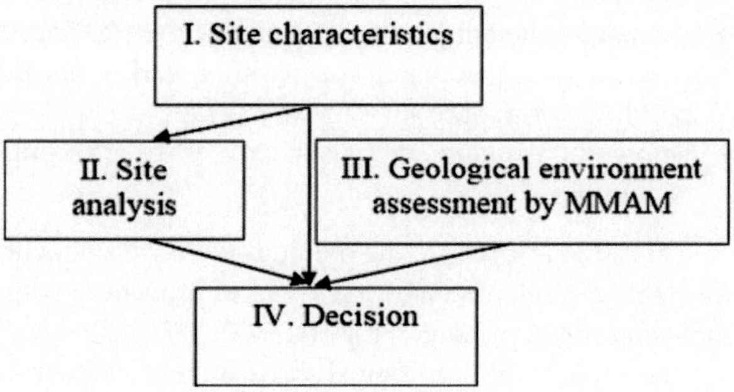

Fig. 4.4 The network of morphological tables for evaluating suitability of construction sites for parking lots. Arrows indicate influence

1. Nearby urban object types (residential buildings; office and administrative buildings; shopping and entertainment centers; stadiums, concert halls, theaters; schools and universities; architectural and tourist attractions; industrial objects);
2. Number of residents in vicinity (up to 1000; 1000–3000; 3000–5000; 5000–10000; over 10000);
3. Number of workplaces in vicinity (up to 500; 500–1000; 1000–3000; 3000–5000; over 5000);
4. Traffic speed (high—over 60 km/h; average—30–60 km/h; low—15–30 km/h; very low—less than 15 km/h);
5. Existing open-air and underground parking places (up to 50; 50–200; 200–400; over 400);
6. Accessibility of the territory for construction (no complications; slight complications; severe complications).

As can be seen, the parameters are characterized by different types of uncertainty—spatial distribution uncertainty (parameter 1—a single site may contain different types of buildings), temporal distribution uncertainty (Parameter 4—the traffic speed may change over time), informational uncertainty (the exact measurements for Parameters 2 and 3 require additional cumbersome research, which is why the expert opinions are deemed sufficient).

For the purposes of the study, "nearby" was defined as a radius of 300 m as a rational assumption for the longest distance that a person would be content to walk from a parked car.

As most of these parameters are independent, the cross-consistency matrix for the table was not introduced. It is implied that the direct estimates for the alternatives would be sufficiently precise and would not require re-adjustment.

The next table "II. Site analysis" contains two parameters that describe the demand for establishing a parking lot on the site—the approximate number of required parking places (Parameter "7. Parking place demand", with the same set of alternatives as Parameter 5 in the previous table), and the type of demand (Parameter "8. Parking demand type"):

- constant—the demand for parking is close to identical at different times;
- pendular—the demand for parking varies at different daytimes (e.g., office buildings or malls);
- peak—the demand emerges in case of some singular mass events (e.g., stadiums, concert halls).

This table is linked to the previous table via a dependency matrix, and the values in this table are calculated by the MMAM procedure using the estimates in the previous table and the dependency matrix.

The table "III. Geological environment assessment by MMAM" contains the parameters from the research of geological environment by MMAM, taken from papers (Pankratova et al 2018; Haiko et al 2019). Only the parameters that directly influence the decision about parking lots, are selected, i.e., "A. Site suitability", "B. Object scale", and "C. Construction depth". In respect to the morphological table network in Fig. 4.4, these parameters were numbered 9–11.

The table "IV. Decision" contains parameters that summarize the decision regarding the considered construction site—the weights for suitability or unsuitability of a site regarding the parking lot construction (parameter "A. Suitability for a parking lot"), and the most advisable sizes of the potential parking lot, taking into account both the demand and the surrounding geological environment (parameter "B. Advisable parking lot size").

The table IV is linked with all the previous tables by the dependence matrix. After the filling of the dependency matrices by expert estimation, the model for evaluating construction sites for a parking lot was obtained.

The model was tested on the same construction sites as in the previous study (Haiko et al. 2019), also mentioned in Sect. 4.3:

- site 1 at the Shevchenkivsky district at the Peremohy avenue;
- site 2 at the Shevchenkivsky district between the Bulvarno-Kudriavska and Honchara streets.

The evaluation of the sites was made by expert estimation, in which the expert was tasked with estimation of each alternative of each parameter in the table "I. Site characteristics" (totaling 28 questions for a single site). The resulting input data for both sites is given in Table 4.31. For each parameter, the weights were normalized so that the sum equaled 1.

The first stage of calculations processed the dependency between the morphological tables "I. Site characteristics" and "II. Site analysis", which resulted in the following estimates for the potential demand in parking places (Table 4.32).

As can be seen, the distribution of demand types (parameter 8) is nearly identical, as for the demand values (parameter 7), site 1 is skewed toward bigger sizes of parking lots compared to site 2.

Table 4.31 The input data for the study, obtained by expert estimation

Parameter	Alternative	Raw input		Normalized input	
		Site 1	*Site 2*	*Site 1*	*Site 2*
1	1.1. Residential buildings	0,8	0,8	0,242	0,184
	1.2. Office and administrative buildings	0,8	0,8	0,242	0,184
	1.3. Shopping and entertainment centers	0,65	0,8	0,197	0,184
	1.4. Stadiums, concert halls, theaters	0,65	0,65	0,197	0,149
	1.5. Schools and universities	0,2	0,65	0,061	0,149
	1.6. Architectural and tourist attractions	0	0,65	0,000	0,149
	1.7. Industrial objects	0,2	0	0,061	0,000
2	2.1. Up to 1000	0	0,2	0,000	0,091
	2.2. 1000–3000	0,35	0,65	0,206	0,295
	2.3. 3000–6000	0,8	0,8	0,471	0,364
	2.4. 6000–10000	0,35	0,35	0,206	0,159
	2.5. Over 10000	0,2	0,2	0,118	0,091
3	3.1. Up to 500	0,35	0,2	0,175	0,100
	3.2. 500–1000	0,8	0,65	0,400	0,325
	3.3. 1000–3000	0,65	0,8	0,325	0,400
	3.4. 3000–5000	0,2	0,35	0,100	0,175
	3.5. Over 5000	0	0	0,000	0,000
4	4.1. High (over 60 km/h)	0,35	0,35	0,206	0,163
	4.2. Average (30…60 km/h)	0,8	0,65	0,471	0,302
	4.3. Low (15…30 km/h)	0,35	0,8	0,206	0,372
	4.4. Very low (less than 15 km/h)	0,2	0,35	0,118	0,163
5	5.1. Up to 50 parking places	0,8	0,8	0,593	0,400
	5.2. 50–200 parking places	0,35	0,65	0,259	0,325
	5.3. 200–400 parking places	0,2	0,35	0,148	0,175
	5.4. Over 400 parking places	0	0,2	0,000	0,100
6	6.1. No complications	0,8	0,65	0,485	0,361
	6.2. Slight complications	0,65	0,8	0,394	0,444
	6.3. Severe complications	0,2	0,35	0,121	0,194

Next, taking the data from Tables 4.31 and 4.32 as input, as well as the results from (Haiko et al. 2019) that describe the geological environment, the decision table is evaluated (Table 4.33).

The obtained results allow to make several conclusions. Both of the sites are quite favorable for construction of parking lots, which can be seen from the estimates of parameter "A. Suitability for a parking lot". This is stipulated by the location of both sites in places with high functional urban activity, close to office and administrative buildings, shopping and entertainment centers, educational facilities, etc. However,

Table 4.32 Parking lot site analysis

Parameter	Alternative	Site 1	Site 2
7	7.1. Up to 50 parking places	0,306	0,379
	7.2. 50–200 parking places	0,407	0,390
	7.3. 200–400 parking places	0,206	0,179
	7.4. Over 400 parking places	0,080	0,053
8	8.1. Constant	0,272	0,275
	8.2. Pendular	0,375	0,372
	8.3. Peak	0,352	0,354

Table 4.33 Decision regarding a parking lot

Parameter	Alternative	Site 1	Site 2
A	A.1. Construction suitable	0,853	0,979
	A.2. Construction unsuitable	0,147	0,021
B	B.1. Up to 50 parking places	0,429	0,355
	B.2. 50–200 parking places	0,444	0,456
	B.3. 200–400 parking places	0,121	0,160
	B.4. Over 400 parking places	0,007	0,028

the second site appeared more favorable, which is explained by its better geological environment assessments, obtained in (Haiko et al. 2019). The most advisable size for potential parking lots was described by the alternative "50–200 parking places" (with weights of 0,444 and 0,456, respectively). High weights of relatively small parking lots can be explained by the quantitative characteristics of the considered sites (bound by the 300 m restriction), as well as the assessments of geological environment which favored lesser scale structures due to their stability (though the impact of this factor can be variable depending on the type of a chosen parking lot).

Thus, the developed technique and tool set allowed to incorporate the assessment of impacts and relations of geologic, technogenic, and structural-functional factors for analysis of favorability of urban territories for underground parking lot construction (taking into account both economic factors and risks of necessity of a parking lot).

4.5 Model of Assessing Potential Underground Tunnel Tracks

The previous studies (Pankratova et al. 2018; Haiko et al. 2019; Pankratova et al. 2019), including the estimation of favorability of city areas for underground construction (Sects. 4.3 and 4.4), allowed to obtain the efficient tools for underground space planning, considering the interaction with engineering geological environment; however, the ecological and safety components were not paid due attention. The construction priorities of the underground objects and complexes, considering the impact on ecological and technical risks, remained an important open issue of the system approach.

The problem of assessing potential car tunnel sites was decomposed into two separate tasks for the two-stage MMAM: the analysis of the geological environment, and the analysis of the structural-functional factors for a studied area.

The analysis of geological factors was conducted using a network of two morphological tables (MT) that was developed by authors earlier (Pankratova et al. 2018), Sect. 4.3, with certain changes aimed at considering the specifics of tunnels as underground objects:

- the impact of the alternative "7.4. Flooded areas/quicksand are present" on the unsuitability of a construction site was significantly increased;
- the ranges of parameters "B. Object scale" and "C. Construction depth" were slightly revised to better reflect the type of analyzed underground objects (i.e., tunnels);
- the set of alternatives for parameter "D. Risk factor" was altered, as a portion of risk factors were transferred to structural and functional factor analysis.

As this study deals with underground construction sites for tunnels, special attention was paid to the alternative "7.4. Flooded areas/quicksand are present". The impact of this alternative on the decision was re-evaluated, as high chances of this alternative significantly increase risks related to construction, and drastically lower the overall favorability of the site.

Testing these changes was made on one of the models of sites in previous studies (Sect. 4.3), with manually increased chance of a flooded area in the generally favorable site. The rest of the probability estimates remained unchanged. Table 4.34 the influence of the new cross-consistency matrix values on the result, specifically the "A. Site favorability" parameter.

Table 4.34 shows how the new model, unlike the old, more adequately reacts to the increase in chances of a present flooded area in the site's geological environment, significantly lowering the "A.1. Suitable" alternative even for a quite favorable from the other points of view site. It should be noted that since the changes to parameter 7 estimates were artificial, an inconsistency between this parameter and the other site parameters appeared, and the model tried to "fix" manual perturbation on the first MMAM stage, slightly shifting them toward the initial values. Even so, the experiment still clearly indicates better adequacy of the new model for the stated task.

The second task was to consider the impact of the structural-functional characteristics of the area, taking into account the risks of ecological and technogenic threats. Firstly, a group of considered technogenic and ecological risks was formed, that could be mitigated by construction of a tunnel:

Table 4.34 Experiment with manual increase of alternative "7.4. Flooded areas/quicksand are present" probability

Low flooded area chance (safe site)				
Estimates of parameter "7. Influence of aquifers and perched groundwater"	Old model		New model	
{0,2; 0,8; 0,2; 0,2}	A.1. Suitable	0,993	A.1. Suitable	0,898
	A.2. Not suitable	0,007	A.2. Not suitable	0,102
Increased flooded area chance (precarious site)				
Estimates of parameter "7. Influence of aquifers and perched groundwater"	Old model		New model	
{0,2; 0,5; 0,5; 0,5}	A.1. Suitable	0,987	A.1. Suitable	0,768
	A.2. Not suitable	0,013	A.2. Not suitable	0,232
High flooded area chance (dangerous site)				
Estimates of parameter "7. Influence of aquifers and perched groundwater"	Old model		New model	
{0,2; 0,2; 0,2; 0,8}	A.1. Suitable	0,981	A.1. Suitable	0,498
	A.2. Not suitable	0,019	A.2. Not suitable	0,502

R1. Air pollution (exhaust emission);
R2. Noise and dynamic impact (engine hum, vehicle clanking, vibration, etc.);
R3. Traffic jams (lowered average traffic speed, disrupting transport communications, increased exhaust emission);
R4. Traffic accidents (injuries, traffic block).

The chosen set of factors conditioned the MMAM application for the task of analyzing structural and functional factors. The MT for the first MMAM stage comprised of eight parameters that were important in regard to the favorability of a planned tunnel and its capacity to mitigate one or more of the risk factors:

1. neighboring urban objects in the planned tunnel's area—this parameter influences the weights of different risk factors. For example, residential buildings, tourist attractions, and parks make mitigating pollution and noise risk factors much more important, compared to industrial objects or undeveloped areas;
2. residential building density—this parameter complements the previous one, influencing risk factor weights;
3. downtown factor—describes the proximity of a studied area to the center of the city, or its influence on the traffic in the city center;
4. crowd density in the planned tunnel's area—this parameter influences the risk factor weights, primarily traffic accidents (more pedestrians mean higher accident

risk), and pollution and noise factors as well. This parameter also impacts the tunnel's capacity to mitigate some of the risk factors, as its construction leads to less opportunities for accidents and traffic jams caused by crosswalks;

5. intensity of traffic in the planned tunnel's area—this parameter is the most critical for the advisability of tunnel construction. Additionally, it influences the risk factor importance—more intensive traffic demands smooth, unobstructed movement, meaning higher weight of the traffic jam risk factor. High traffic intensity also increases the tunnel's capacity of dealing with all the risk factor groups: rerouting more intensive traffic underground means stronger mitigation of all the considered risk factors;
6. average traffic speed on the busiest road sections at peak hours in the planned tunnel's area—this parameter varies the tunnel's influence on risk factors: low movement speed means that the tunnel's construction will strongly impact the pollution and traffic jam factors, while high movement speed means the same for noise and traffic accident factors. Higher movement speed also increases the weight of traffic accident factor, as the accidents become potentially more dangerous;
7. surface connectivity of the planned tunnel ends by existing roads—the parameter obviously determines the advisability of the tunnel's construction. Also, this parameter provides weight to the traffic jam factor, as under bad connectivity the presence or absence of traffic jams becomes critical. Accordingly, there is an opposite effect—if the connectivity is poor, the tunnel's construction creates a positive effect on traffic by creating an alternative route;
8. surface road throughput in the planned tunnel's area (road width, presence of crossings, especially unregulated crossing)—the parameter influences the advisability of the tunnel and the traffic jam factor weight. It also contributes to tunnel's capacity of mitigating traffic jam and accident risk factors.

At the second MMAM stage an MT was constructed that describes the advisability of tunnel's construction from the structural-functional point of view, as well as the profile of the planned tunnel's area from the ecological and safety risk factors point of view. Five parameters in the table were assigned to risk analysis: one parameter compares the weights of the different risk factors, describing an "image" of the area in regard to the risk factor importance; the other four parameters indicate the tunnel's capacity of mitigating each of the considered risk factor groups R1–R4. The MT consists of these parameters:

A. Tunnel construction advisability (advisable, not advisable);
B. Risk factor importance in the planned tunnel's area (air pollution; noise and dynamic impact; traffic jams; traffic accidents);
C. Impact of tunnel construction at the pollution risk factor (no impact; slightly mitigates; moderately mitigates; significantly mitigates);
D. Impact of tunnel construction at the noise and dynamic impact risk factor (no impact; slightly mitigates; moderately mitigates; significantly mitigates);
E. Impact of tunnel construction at the traffic jams risk factor (no impact; slightly mitigates; moderately mitigates; significantly mitigates);

F. Impact of tunnel construction at the traffic accidents risk factor (no impact; slightly mitigates; moderately mitigates; significantly mitigates).

After the expert estimation, the MMAM procedure was utilized to obtain the weights for alternatives of the second-stage MTs, considering the possibility of emergence of any configuration at the first stage with the respective probability, calculated at the first MMAM stage. The number of potential configurations equaled 524288 for the first task and 41472 for the second task.

Testing on the example of car tunnels in Kyiv center. The General Plan of Kyiv city development up to 2025 envisions the construction of 8 car tunnels, three of which will pass under the Dnipro River (Fig. 4.5). These decisions regarding Kyiv's infrastructure development are not only aimed at solving logistical problems but should also improve the ecological situation, as ecologized tunnels provide the opportunity for purposeful redirection and utilization of noxious exhaust fumes from transport (Cuia 2019; Haiko et al. 2016). It is important to analyze the planned tunnel tracks and determine the priorities of tunnel construction, using the ecological safety criterion, to select those that provide the maximum mitigation of ecological risks.

Fig. 4.5 The scheme of car tunnel tracks (the General Plan of Kyiv city development up to 2025)

Judging by the necessity to address the most urgent problems of transport situation in Kyiv, the morphological models for two car tunnels of the right bank Kyiv were developed: Tunnel 1 (M. Gryshko National Botanical Garden—the Darnytskyi Bridge), and Tunnel 5 (Peremohy square—the Dnipro River; only the section before Dnipro was considered, without the underwater tunnel section).

The results of expert assessment of the geological environment and the values calculated by the first MMAM stage are presented in Table 4.35.

Column "1" in Table 4.35 shows the normalized expert assessments, and Column "2" shows the processed by MMAM values which take into account the mutual dependencies between parameters. As the table shows, the influence of interdependence allowed to obtain a more precise representation; however, significant differences in distributions between alternatives were not observed, thus affirming that the expert assessment was conducted quite accurately, and the parameter estimates are consistent.

Using the calculated values, the second MMAM stage is implemented to evaluate decisions regarding the geological engineering factors (Table 4.36).

Table 4.36 implies that the geological environments around both tunnel tracks are favorable enough for underground construction (the weight of "A.1. Suitable" alternative is considerably higher than the weight of "A.2. Not suitable" alternative), and the difference between this parameter for both tunnels is negligible. According to the characteristics of Tunnel 5 (shown in Table 1), the model proposes smaller cross-section and deeper construction depth compared to Tunnel 1 ("C.4. beneath 60 m" weight is 0,518 for Tunnel 5 and 0,276 for Tunnel 1), which can be explained by land relief. The two considered sites also have slightly different profiles regarding risk factors—the most tangible risk factor for Tunnel 1 is "D.4. Initiating displacements and other unwanted geological processes" with weight 0,547, which is significantly higher than the next significant risk factor "D.2. Increasing construction and operation cost" with weight 0,348, whereas, for Tunnel 5, both of these factors are nearly equivalent (their weights are 0,426 and 0,417, respectively). The profiles for risk degree (probability) and risk level (cost of consequences) are close for both tunnels, with a slight skew toward higher values for Tunnel 5 (Figs. 4.6 and 4.7), which means that the cost of support and potential renovations for Tunnel 5 will be 5–7% of total construction value more compared to Tunnel 1. Therefore, the geological environments, that are favorable enough for both of the tunnels, cannot be the defining factor for construction priority.

The next step was to assess the structural-functional factors by the second morphological table (Table 4.37). Here the input estimates are given (the "Input" column), as well as the values calculated by MMAM, using two alternative sets of model coefficients by different experts (columns labeled "Exp1" and "Exp2").

The results of this estimation allow to make a comparison of the two areas:

Table 4.35 The geological environment assessments for two tunnel tracks

Parameter	Parameter alternatives	Tunnel 1		Tunnel 5	
		1	2	1	2
1. Level of dynamic load	1.1. Low (46–53 dB)	0,143	0,137	0,093	0,070
	1.2. Medium (53–73 dB)	0,327	0,489	0,302	0,448
	1.3. Increased (73–96 dB)	0,327	0,238	0,372	0,299
	1.4. High (over 96 dB)	0,204	0,136	0,233	0,182
2. Static load from surface buildings	2.1. Insignificant (Ksl < 1)	0,206	0,101	0,000	0,000
	2.2. Medium (1 < Ksl < 2)	0,471	0,584	0,194	0,164
	2.3. Increased (2 < Ksl < 3,5)	0,206	0,215	0,444	0,451
	2.4. High (Ksl > 3,5)	0,118	0,100	0,361	0,385
3. Static load from soil	3.1. Insignificant (Kmas < 0,05, MPa)	0,175	0,063	0,093	0,015
	3.2. Medium (0,05 < Kmas < 0,3, MPa)	0,400	0,703	0,302	0,485
	3.3. High (0,3 < Kmas < 0,5, MPa)	0,325	0,200	0,372	0,362
	3.4. Very high (Kmas > 5, MPa)	0,100	0,034	0,233	0,138
4. Influence of existing underground objects	4.1. Absent (distance over 50 m)	0,129	0,055	0,108	0,032
	4.2. Slight (distance 20–50 m)	0,516	0,422	0,432	0,336
	4.3 Significant (distance 10–20 m)	0,226	0,437	0,351	0,584
	4.4 Hazardous (distance less than 10 m)	0,129	0,087	0,108	0,048
5. Genetic type and lithologic composure of soil	5.1. Unweathered clays and average density sands	0,175	0,269	0,265	0,320
	5.2. Technogenic deposits (alluvial and bulk types)	0,100	0,127	0,143	0,157
	5.3. Deluvial clay soils (water-saturated), water-saturated overfloodplain sands	0,400	0,402	0,327	0,354
	5.4. Sedentary soils, soils with special properties (loess, peat, silt)	0,325	0,203	0,265	0,169
6. Effective soil strength	6.1. Very strong soils > 300 kPa	0,000	0,000	0,093	0,091
	6.2. Strong soils 200–300 kPa	0,121	0,113	0,233	0,328

(continued)

Table 4.35 (continued)

Parameter	Parameter alternatives	Tunnel 1		Tunnel 5	
		1	2	1	2
	6.3. Average strength soils 150–200 kPa	0,485	0,575	0,372	0,402
	6.4. Relatively strong soils < 150 kPa	0,394	0,313	0,302	0,179
7. Influence of aquifers and perched groundwater	7.1. Water-bearing horizons at P-N1np	0,217	0,193	0,372	0,383
	7.2. Groundwater depth > 3 m, pressurized groundwater > 10 m	0,283	0,365	0,372	0,474
	7.3. Groundwater depth < 3 m, pressurized groundwater < 10 m	0,348	0,358	0,093	0,089
	7.4. Flooded areas/quicksand are present	0,152	0,084	0,163	0,054
8. Landscape type and morphometrics	8.1. Flat areas of overfloodplain terraces, morainic-glacial plains	0,143	0,159	0,000	0,000
	8.2. Slightly tilted overfloodplain terraces, watershed areas	0,265	0,340	0,194	0,353
	8.3. Small river valleys, slightly irregular slopes, high floodplain	0,327	0,416	0,361	0,536
	8.4. Slope areas with ravines and steep banks, low floodplain	0,265	0,086	0,444	0,111
9. Geological engineering processes	9.1. Absent	0,082	0,043	0,093	0,027
	9.2. Stabilized	0,265	0,320	0,302	0,303
	9.3. Low displacement processes	0,327	0,527	0,372	0,586
	9.4. Active manifestations of subsidence, underflooding, gravitational processes	0,327	0,109	0,233	0,084
10. Geo-technologies	10.1. Open	0,200	0,364	0,200	0,119
	10.2. Underground	0,800	0,636	0,800	0,881

- the area of Tunnel 5 is mostly covered in residential, commercial, and administrative buildings, while a large portion of Tunnel 1 area is taken by parks and undeveloped areas. Accordingly, Tunnel 5 has much higher density of residential buildings (Parameter 2): the highest-ranking alternatives are "High" and "Average", whereas for Tunnel 1 they are "Average" and "Low";

Table 4.36 Estimates of the decision alternative's weights obtained at the second MMAM stage

Parameter	Alternative	Tunnel 1	Tunnel 5
A. Track suitability	A.1. Suitable	0,777	0,799
	A.2. Not suitable	0,223	0,201
B. Object scale	B.1. Cross-section up to 10 m^2	0,643	0,712
	B.2. Cross-section up to 25 m^2	0,255	0,232
	B.3. Cross-section up to 40 m^2	0,084	0,049
	B.4. Cross-section up to and over 40 m^2	0,018	0,007
C. Construction depth	C.1. 0–10 m	0,096	0,021
	C.2. 10–20 m	0,183	0,100
	C.3. 20–50 m	0,445	0,360
	C.4. beneath 60 m	0,276	0,518
D. Risk factor	D.1. Construction failure, malfunction	0,039	0,046
	D.2. Increasing construction and operation cost	0,348	0,417
	D.3. Dangerous influence on surface or neighboring underground objects	0,067	0,111
	D.4. Initiating displacements and other unwanted geological processes	0,547	0,426
E. Risk degree	E.1. <3%	0,089	0,050
	E.2. 3–10%	0,598	0,584
	E.3. 10–20%	0,274	0,308
	E.4. 20–50%	0,031	0,044
	E.5. > 50%	0,009	0,014
F. Risk level	F.1. 0,1–5% Q	0,212	0,137
	F.2. 5–20% Q	0,701	0,768
	F.3. 20–50% Q	0,079	0,085
	F.4. >50% Q	0,008	0,010

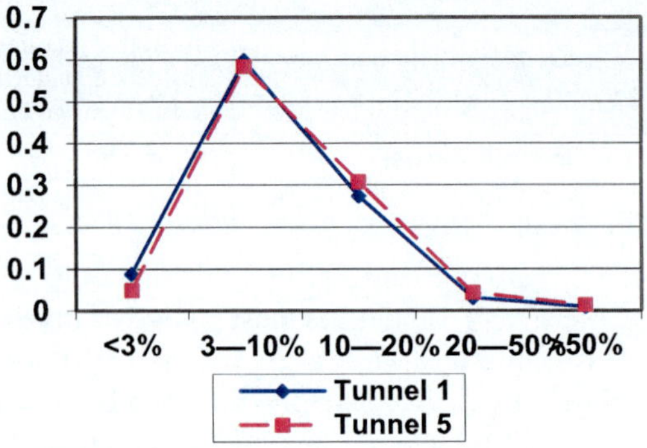

Fig. 4.6 The estimates of degree of risk for Tunnels 1 and 5 (Q is the total tunnel construction cost)

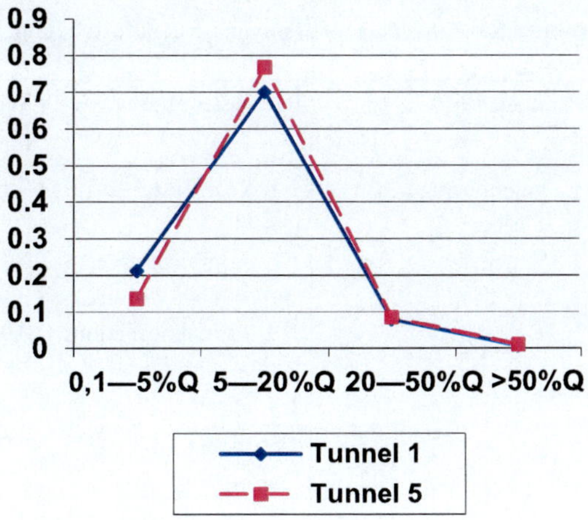

Fig. 4.7 The estimates of level of risk for Tunnels 1 and 5 (Q is the total tunnel construction cost)

- the area of Tunnel 5 can undoubtedly be labeled as historical city center zone, and it definitely influences the traffic in the city center; not so much for the Tunnel 1 area, which is mostly outside the historical city center;
- the crowd density in the Tunnel 5 area is average to high (due to few high-rise residential buildings), and in the Tunnel 1 area the crowd density is low to average (due to average building density, and the presence of park areas);
- intensity of traffic is higher in the Tunnel 5 area; however, at peak hours, the traffic speed is lower in comparison with the Tunnel 1 area;
- the surface road connectivity and throughput are better for the Tunnel 1 area.

Using the estimates obtained in Table 4.37, and the constructed dependency matrix, the second MMAM stage was conducted for assessing the impact of structural-functional factors. This calculation was also performed at two alternative sets of model coefficients (columns labeled "Exp1" and "Exp2"), the results are given in Table 4.38.

For the sake of convenience, the average estimates of two models are referenced. The results of evaluation allow to make several conclusions regarding the planned tunnels' sites:

1. The construction of both tunnels is highly advisable: the weight of alternative "A.1. Advisable" significantly overcomes the weight of "A.2. Not advisable" (accordingly, 0,895 versus 0,105 for Tunnel 1, and 0,994 versus 0,006 for Tunnel 5). This is a very reasonable result, as the test dealt with the real, previously justified objects from the General Kyiv city plan. The integral advisability factor for Tunnel 5 emerged 10% higher than for Tunnel 1.
2. The structure of the most influential risk factors has slight differences for the considered tunnel areas. The calculation results for parameter "B. Risk factor importance in the planned tunnel's area" are presented in Fig. 4.8. The biggest difference is that the traffic jams risk factor is more important in the Tunnel 5

Table 4.37 Input data for assessing structural-functional factors, and the results of the first MMAM stage

Parameter	Parameter alternative	Tunnel 1			Tunnel 5		
		Input	Exp1	Exp2	Input	Exp1	Exp2
1. Neighboring urban objects	1.1. Residential buildings	0,197	0,250	0,328	0,258	0,471	0,315
	1.2. Administrative, commercial buildings	0,197	0,194	0,188	0,210	0,298	0,272
	1.3. Tourist attractions, sights	0,061	0,043	0,068	0,258	0,176	0,261
	1.4. Parks	0,242	0,219	0,320	0,210	0,053	0,152
	1.5. Industrial buildings	0,061	0,055	0,069	0,000	0,000	0,000
	1.6. Undeveloped areas	0,242	0,240	0,028	0,065	0,002	0,000
2. Residential building density	2.1. Very low	0,175	0,130	0,046	0,000	0,000	0,000
	2.2. Low	0,400	0,450	0,100	0,121	0,039	0,000
	2.3. Average	0,325	0,332	0,704	0,485	0,488	0,575
	2.4. High	0,100	0,088	0,150	0,394	0,473	0,425
3. Downtown factor	3.1. Area is in the city center	0,121	0,089	0,285	0,500	0,455	0,845
	3.2. Area influences traffic in the city center	0,394	0,401	0,164	0,500	0,545	0,155
	3.3. Area is far from city center	0,485	0,510	0,552	0,000	0,000	0,000
4. Crowd density	4.1. Very low	0,233	0,152	0,054	0,100	0,007	0,001
	4.2. Low	0,372	0,367	0,143	0,100	0,029	0,008
	4.3. Average	0,302	0,381	0,598	0,400	0,414	0,427
	4.4. High	0,093	0,100	0,205	0,400	0,550	0,565
5. Intensity of traffic	5.1. Low	0,093	0,165	0,046	0,000	0,000	0,000
	5.2. Average	0,302	0,287	0,576	0,289	0,151	0,251
	5.3. High	0,372	0,390	0,345	0,356	0,418	0,595
	5.4. Very high	0,233	0,157	0,034	0,356	0,431	0,154
6. Average traffic speed at peak hours	6.1. Lower than 15 km/h	0,148	0,156	0,033	0,194	0,297	0,159
	6.2. 15–30 km/h	0,259	0,385	0,229	0,444	0,615	0,658
	6.3. 30–60 km/h	0,593	0,459	0,738	0,361	0,088	0,183
7. Surface connectivity	7.1. Very poor	0,129	0,110	0,036	0,325	0,216	0,243
	7.2. Poor	0,516	0,462	0,358	0,400	0,318	0,545
	7.3. Average	0,226	0,246	0,326	0,175	0,252	0,134
	7.4. Good	0,129	0,182	0,281	0,100	0,214	0,078

(continued)

Table 4.37 (continued)

Parameter	Parameter alternative	Tunnel 1			Tunnel 5		
		Input	Expl	Exp2	Input	Expl	Exp2
8. Surface road throughput	8.1. Low	0,233	0,211	0,119	0,361	0,367	0,536
	8.2. Average	0,533	0,540	0,648	0,444	0,513	0,432
	8.3. High	0,233	0,250	0,233	0,194	0,120	0,033

Table 4.38 Assessing impact of structural-functional factors at tunnel construction areas on the ecological and safety factors

Parameter	Parameter alternative	Tunnel 1		Tunnel 5	
		Expl	Exp2	Expl	Exp2
A. Tunnel construction advisability	A.1. Advisable	0,897	0,893	0,988	0,999
	A.2. Not advisable	0,103	0,107	0,012	0,001
B. Risk factor importance in the planned tunnel's area	B.1. Air pollution	0,151	0,134	0,182	0,164
	B.2. Noise and dynamic impact	0,127	0,205	0,137	0,165
	B.3. Traffic jams	0,445	0,115	0,527	0,380
	B.4. Traffic accidents	0,277	0,546	0,154	0,291
C. Impact of tunnel construction at the pollution risk factor	C.1. No impact	0,134	0,042	0,033	0,010
	C.2. Slightly mitigates	0,341	0,466	0,291	0,287
	C.3. Moderately mitigates	0,322	0,300	0,389	0,327
	C.4. Significantly mitigates	0,204	0,193	0,287	0,376
D. Impact of tunnel construction at the noise and dynamic impact risk factor	D.1. No impact	0,092	0,080	0,091	0,020
	D.2. Slightly mitigates	0,317	0,306	0,295	0,244
	D.3. Moderately mitigates	0,374	0,349	0,380	0,361
	D.4. Significantly mitigates	0,217	0,265	0,233	0,375
E. Impact of tunnel construction at the traffic jams risk factor	E.1. No impact	0,030	0,185	0,006	0,013
	E.2. Slightly mitigates	0,398	0,486	0,280	0,193
	E.3. Moderately mitigates	0,395	0,174	0,464	0,339
	E.4. Significantly mitigates	0,177	0,154	0,249	0,455
F. Impact of tunnel construction at the traffic accidents risk factor	F.1. No impact	0,090	0,079	0,038	0,007
	F.2. Slightly mitigates	0,315	0,367	0,315	0,438
	F.3. Moderately mitigates	0,357	0,339	0,388	0,330
	F.4. Significantly mitigates	0,238	0,214	0,259	0,225

area, while the same is true for the traffic accident risk factor in the Tunnel 1 area, which can be explained by higher traffic speed. Also, the air pollution risk factor has more weight for Tunnel 5 area.

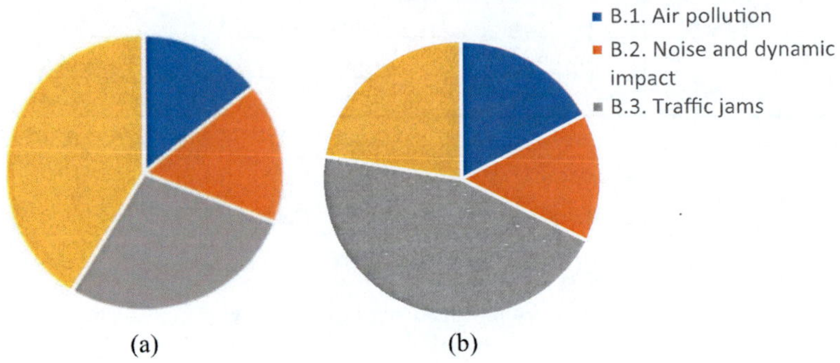

Fig. 4.8 Diagrams for risk factor weights at the areas of Tunnel 1 (**a**) and Tunnel 5 (**b**)

3. The construction of each of the studied tunnels will definitely provide a certain mitigation of the considered risk factors, and the extent of this mitigation varies for each specific tunnel and risk factor type.

The estimated impact on all the considered risk factors for both tunnels is shown in Figs. 4.9, 4.10, 4.11 and 4.12.

Studying the diagrams presented in Figs. 4.9, 4.10, 4.11 and 4.12, and the values from Table 4.38, several conclusions can be made.

The weight of alternative "No impact" is notably small, especially for Tunnel 5, proving the high potential of underground construction for ecologization of transport infrastructure. This fact is also confirmed by high values of the alternative "A.1. Tunnel advisability", as was noted earlier.

In general, Tunnel 5 provides higher mitigation of ecological and safety risks of existing surface infrastructure. For the factors "R2. Noise and dynamic impact" and "R4. Traffic accidents" its advantage is negligible (within the estimation error margin), however, for the more important factors "R1. Air pollution" and "R3. Traffic

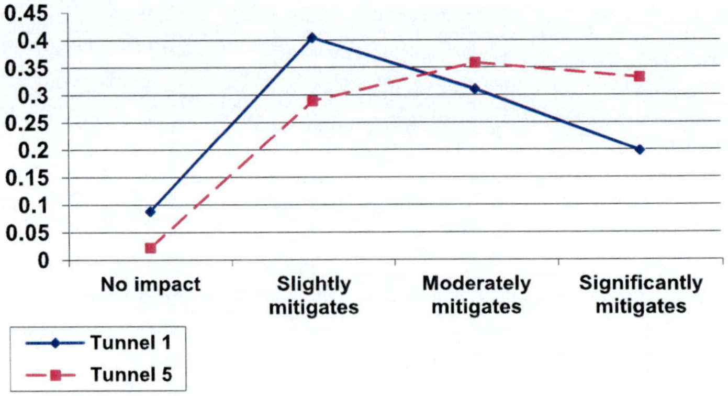

Fig. 4.9 Diagram of tunnels' impact on the risk factor "R1. Air pollution"

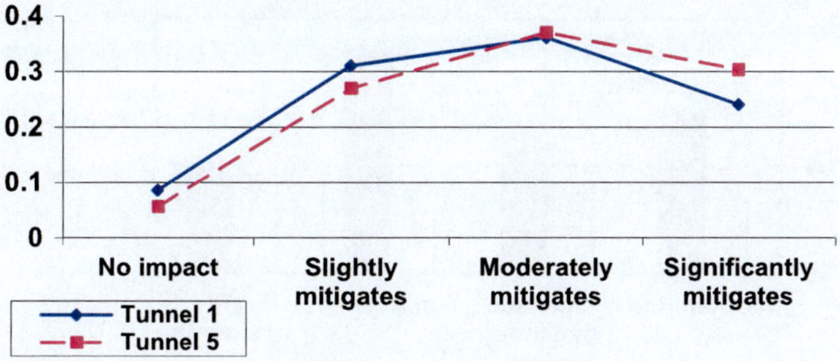

Fig. 4.10 Diagram of tunnels' impact on the risk factor "R2. Noise and dynamic impact"

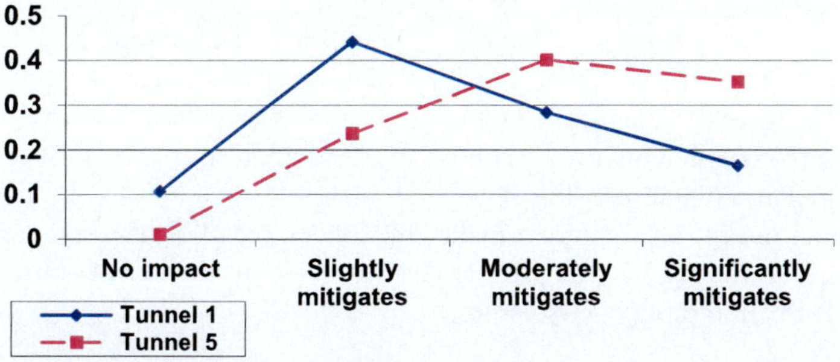

Fig. 4.11 Diagram of tunnels' impact on the risk factor "R3. Traffic jams"

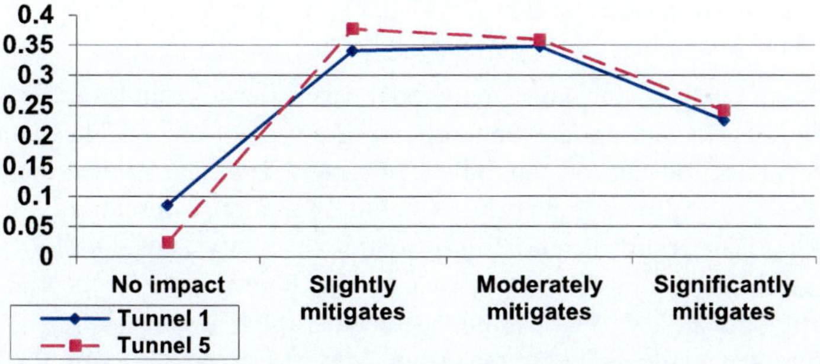

Fig. 4.12 Diagram of tunnels' impact on the risk factor "R4. Traffic accidents"

jams" the advantage of Tunnel 5 is more evident. While the impact of Tunnel 1 can be characterized as slight to moderate, Tunnel 5 shifts the impact to the moderate to significant zone.

To make comparisons of tunnel construction impact on risk factors more demonstrative, let us introduce the impact weight coefficient: $w_{\text{ratio}} = (w_{\text{medium}} + w_{\text{high}})/(w_{\text{low}} + w_{\text{none}})$, that denotes the ratio between the sum of moderate w_{medium}

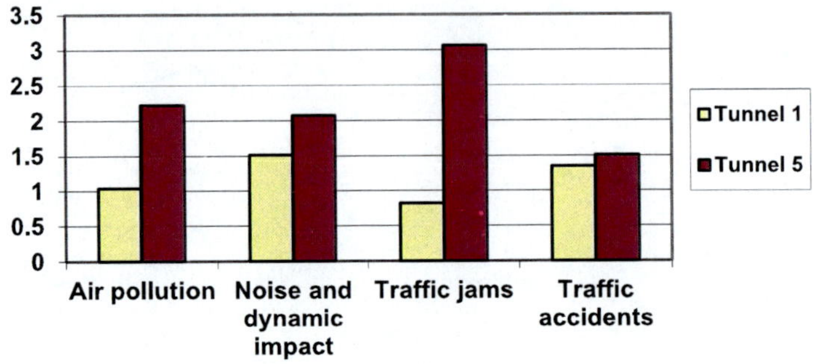

Fig. 4.13 Impact weight coefficient w_{ratio} of Tunnels 1 and 5 for different risk factors

and significant w_{high} impact to the sum of slight w_{low} and absent w_{none} impact on mitigation of a given risk factor. Higher values of this coefficient represent more significant influence of the planned tunnel's construction on the respective risk factor. The calculated coefficients w_{ratio} for both of the tunnels and each of the risk factors are presented as a diagram in Fig. 4.13.

The given in Fig. 4.13 diagram makes the advantage of Tunnel 5 more obvious, making its construction the first priority task for quick improvement of transport and ecological situation in Kyiv city center.

4.6 Morphological Analysis of Undesirable Events for Urban Underground Objects

The purpose of the next morphological model was to describe undesirable events that can potentially affect an underground object or a type of object. These undesirable events encompass both natural phenomena, disasters, and catastrophes, and the events of technogenic or anthropogenic origin (including malicious intent: military actions, terrorist acts). The result of modeling provides the analysis of consequences from an undesirable event, allowing to compare several objects or projects on the basis of their stability and ability to withstand various disruptive events.

Modeling was made using the two-stage MMAM procedure, with the first stage describing the multitude of undesirable events, and the second stage analyzing the consequences on the studied object from different angles. A peculiarity of this research is that different objects and types of objects often have fundamentally different interactions between the characteristics of an undesirable event and its consequences. This means that each individual object requires not only the preliminary input values for alternatives but also its own cross-consistency and dependency matrix values.

To select the main characteristic parameters of undesirable events and their consequences, a number of legislative and normative documents were analyzed. As a result, three main characteristic parameters of undesirable events relevant to the study were selected:

Parameter 1: Type of undesirable event. Only the prime cause, or the trigger of an undesirable event, is considered. Of course, these events often cause chain reactions: for example, an explosion leads to a fire, which leads to destruction, etc.; however, in this study, the secondary damage is viewed as consequences of the primary event, because all possible variants of chains of undesirable events are impossible and impractical to consider.

Six alternatives were chosen as types of undesirable events:

- explosion;
- fire;
- landslides, landfalls, subsidence of soil;
- weather cataclysms;
- operational damage and/or destruction of object or its parts;
- functional interruption without destruction.

This study was aimed at creating a universal model, which is why some of the undesirable events are impossible in principle (e.g., a fire for an underwater drainage pipe). In these cases, an impossible alternative receives the value "0" and, accordingly, has no impact on the further algorithms of MMAM.

Parameter 2: Origin of undesirable event. Here 4 alternative origins of an undesirable event were selected:

- anthropogenic with malicious intent (terrorism, diversions, military actions);
- anthropogenic without malicious intent (errors, negligence, non-compliance to construction and exploitation safety);
- technical, technological (technical failures, breakdowns, destruction caused by technological factors, corrosion, etc.);
- natural (atmospheric, hydrospheric, lithospheric disturbance, and natural disasters).

Parameter 3: Scale of undesirable event's impact. Here five alternatives are considered:

- separate structural or functional element, or limited area;
- several structural or functional elements, or several areas;
- the whole object;
- the whole object, and neighboring objects;
- city region and more.

Larger catastrophes are beyond the scope of this research, since it considers only the consequences for a specific urban object. Distinguishing between larger catastrophes is unnecessary for the study, so they were combined into this single alternative.

The morphological table based on these parameters is shown in Table 4.39.

It is quite evident that the parameters of this table are interrelated, so the cross-consistency matrix should be assessed here; moreover, different types of objects may have different dependencies between objects, requiring to evaluate this matrix individually for each type of a studied object.

The morphological table at the second stage contains the parameters and alternatives of consequences of undesirable events. Since these consequences are quite diverse, their comprehensive analysis required 8 parameters.

Parameter A. Integrity of the object and its parts. This parameter describes the damage caused by an undesirable event. Four fundamentally different alternatives were chosen:

- no damage or negligible damage;
- damage may be undone without interruption of operation;
- damage may be undone with interruption of operation;
- damage is irreversible.

Parameter B. Operational capacity. This parameter describes the object's capacity to perform its functions under the influence of an undesirable event. Three alternatives were chosen:

Table 4.39 Description of undesirable events

Parameter	Alternative
1. Type of undesirable event	1.1 Explosion
	1.2 Fire
	1.3 Landslides, landfalls, subsidence of soil
	1.4 Weather cataclysms
	1.5 Operational damage and/or destruction of object or its parts
	1.6 Functional interruption without destruction
2. Origin of undesirable event	2.1 Anthropogenic with malicious intent
	2.2 Anthropogenic without malicious intent
	2.3 Technical, technological
	2.4 Natural
3. Scale of undesirable event's impact	3.1 Separate structural or functional element, or limited area
	3.2 Several structural or functional elements, or several areas
	3.3 Whole object
	3.4 Whole object and neighboring objects
	3.5 City region and more

- object may perform all of its functions;
- object may perform a portion of its functions;
- object stops functioning.

Parameter C. Potential to transfer functions to other objects. This parameter defines if the object's functions can be transferred to other urban objects in case of emergency (i.e., the incapacity to perform its own functions as a result of an undesirable event). Four alternatives were chosen:

- object's functions can be transferred without limitations;
- object's functions can be transferred with some limitations;
- object's functions can be transferred with significant limitations;
- object's functions cannot be transferred.

Parameter D. Operation restore time. This parameter describes the necessary time period to restore the object's operation after the impact of an undesirable event. Five alternatives were chosen:

- operation restore time is unnecessary;
- operation restore time up to 7 days;
- operation restore time up to 1 month;
- operation restore time up to 1 year;
- the object cannot be restored during 1 year.

Parameter E. Casualties. This parameter indicates the approximate number of injured persons in case an undesirable event happens. According to the normative documents, five alternatives were defined:

- none;
- up to 10 persons;
- 10–50 persons;
- 50–200 persons;
- more than 200 persons.

Parameter F. Affected citizens. This parameter indicates the approximate number of people who have their living conditions disrupted in case an undesirable event happens. According to the normative documents, five alternatives were defined:

- none;
- up to 10 persons;
- 10–100 persons;
- 100–1000 persons;
- more than 1000 persons.

Parameter G. Material damage. This parameter indicates the approximate value of material damage in case an undesirable event happens at the object. The material damage is assessed relative to the minimum wage value (MW). Four alternatives were defined according to the normative documents:

- up to 100 MW;
- 100–1000 MW;
- 1000–10000 MW;
- more than 10000 MW.

Parameter H. Ecological consequences. This parameter describes the potential risks for the ecological situation in case an undesirable event happens at the object. Four alternatives were chosen:

- no tangible ecological consequences;
- slight, local, short-term worsening of the ecological situation;
- significant long-term worsening of the ecological situation in a large area;
- ecological catastrophe.

Parameters of the undesirable event consequences are presented in the form of a morphological Table 4.40.

The model was implemented for two chosen critical urban infrastructure objects: (a) a complex of underwater pipes, and (b) a potential underground tunnel for pipes under the Dnipro River, both intended for sewage system.

Using the two-stage MMAM procedure, the analysis of consequences from a multitude of possible undesirable events was made. The results of calculation are given in Table 4.41. The results for a complex of underwater pipes are given in the column "Pipes", and the results for an underground tunnel are given in the column "Tunnel".

Table 4.41 allows to make several comparative conclusions:

- generally, an underwater tunnel provides for better resistance to potential damage in case of any undesirable events. Parameter A (Integrity of the object and its parts) has the same most probable alternative A.3—"Damage may be undone with interruption of operation" for both objects (with weights 0,544 for underwater pipes, and 0,530 for underground tunnel); however, the second most significant alternative is A.4—"Damage is irreversible" for underwater pipes (with 0,416 weight), while for underground tunnel the same is true for alternative A.2—"Damage may be undone without interruption of operation" (with 0,417 weight), and the weight of A.4—"Damage is irreversible" is close to zero for an underground tunnel. This situation is even more demonstrative for parameter B (Operational capacity): underwater pipes have the weight 0,981 of B.3—"Object stops functioning", pointing at very low resistance to damage in case of undesirable events. For comparison, the weight of the same alternative for an underground tunnel is 0,075, meaning that it is highly persistent to total cease of its operation;
- when considering parameter C (Potential to transfer functions to other objects) it is worth to note that, in the studied concept of the underground tunnel for pipes, the existing system of pipes is not dismantled, but left as a reserve system, which can explain the weights received by alternatives of this parameter for the underground tunnel: C.2—"Object's functions can be transferred with some limitations" has value 0,561, and C.3—"C.3 Object's functions can be transferred with significant limitations" with value 0,401. Underwater pipes have the largest weights for

Table 4.40 Description of undesirable event consequences

Parameter	Alternative
A. Integrity of the object and its parts	A.1 No damage or negligible damage
	A.2 Damage may be undone without interruption of operation
	A.3 Damage may be undone with interruption of operation
	A.4 Damage is irreversible
B. Operational capacity	B.1 Object may perform all of its functions
	B.2 Object may perform a portion of its functions
	B.3 Object stops functioning
C. Potential to transfer functions to other objects	C.1 Object's functions can be transferred without limitations
	C.2 Object's functions can be transferred with some limitations
	C.3 Object's functions can be transferred with significant limitations
	C.4 Object's functions cannot be transferred
D. Operation restore time	D.1 Operation restore time is unnecessary
	D.2 Operation restore time up to 7 days
	D.3 Operation restore time up to 1 month
	D.4 Operation restore time up to 1 year
	D.5 The object cannot be restored during 1 year
E. Casualties	E.1 None
	E.2 Up to 10 persons
	E.3 10–50 persons
	E.4 50–200 persons
	E.5 More than 200 persons
F. Affected citizens	F.1 None
	F.2 Up to 10 persons
	F.3 10–100 persons
	F.4 100–1000 persons
	F.5 More than 1000 persons
G. Material damage	G.1 Up to 100 MW
	G.2 100–1000 MW
	G.3 1000–10000 MW
	G.4 More than 10000 MW
H. Ecological consequences	H.1 No tangible ecological consequences

(continued)

Table 4.40 (continued)

Parameter	Alternative
	H.2 Slight, local, short-term worsening of the ecological situation
	H.3 Significant long-term worsening of the ecological situation in a large area
	H.4 Ecological catastrophe

alternatives C.4—"Object's functions cannot be transferred" (value 0,574), and C.3—"Object's functions can be transferred with significant limitations" (value 0,322);

- parameter D (Operation restore time) also shows advantage of underground tunnel over underwater pipes. The alternatives with the most weight are D.4—"Operation restore time up to 1 year" (value 0,507) and D.5—"The object cannot be restored during 1 year" (value 0,246) for underwater pipes. As for the underground tunnel, its alternatives with the most weight are D.3—"Operation restore time up to 1 month" (value 0,524) and D.2—"Operation restore time up to 7 days" (value 0,455).
- similar results were obtained for parameter G (Material damage). Underwater pipes have the following ranking of alternatives: G.3—"1000–10000 MW" (value 0,431), G.4—"More than 10000 MW" (value 0,309), G.2—"100–1000 MW" (value 0,259), and the underground tunnel has the following ranking: G.2—"100–1000 MW" (value 0,512), G.1—"Up to 100 MW" (value 0,464), meaning that the process of restoring an underground tunnel generally takes up to 10 times less resources compared to the underwater pipes;
- Parameter E (Casualties) is not tangible in this study due to the nature of the considered objects. Direct casualties are close to impossible, since the process of transferring sewage is mostly automated, without human presence. The importance of this parameter will be more significant for other types of urban objects;
- the estimation results for parameter F (Affected citizens) again prove the results obtained for previous parameters. Since the operation will most likely be disrupted in case an undesirable event happens to underwater pipes, the affected urban population will be very high (F.5—"More than 1000 persons", with weight 0,551). An underground tunnel received the highest weight for alternative F.1—"None", with a weight of 0,811. Intermediate alternatives F.2—"Up to 10 persons", F.3—"10–100 persons" in both cases received very low values, since disrupting the sewage system immediately causes harm to living conditions of a large number of people, underlining the critical nature of this urban infrastructure element;
- the parameter H (Ecological consequences) is the most convincing when showing the advantage of a sewage system in an underground tunnel compared to underwater pipes, as the ecological consequences in case an undesirable event happens are mostly negligible for an underwater tunnel (alternative H.1—"No tangible ecological consequences" with weight 0,894), while disruptions for underwater

Table 4.41 Estimation of undesirable events consequences

Parameter	Alternative	Estimate	
		Pipes	Tunnel
A. Integrity of the object and its parts	A.1 No damage or negligible damage	0,018	0,051
	A.2 Damage may be undone without interruption of operation	0,021	0,417
	A.3 Damage may be undone with interruption of operation	0,544	0,530
	A.4 Damage is irreversible	0,416	0,001
B. Operational capacity	B.1 Object may perform all of its functions	0,000	0,269
	B.2 Object may perform a portion of its functions	0,019	0,656
	B.3 Object stops functioning	0,981	0,075
C. Potential to transfer functions to other objects	C.1 Object's functions can be transferred without limitations	0,005	0,025
	C.2 Object's functions can be transferred with some limitations	0,100	0,561
	C.3 Object's functions can be transferred with significant limitations	0,322	0,401
	C.4 Object's functions cannot be transferred	0,574	0,014
D. Operation restore time	D.1 Operation restore time is unnecessary	0,000	0,004
	D.2 Operation restore time up to 7 days	0,009	0,455
	D.3 Operation restore time up to 1 month	0,174	0,524
	D.4 Operation restore time up to 1 year	0,570	0,017
	D.5 The object cannot be restored during 1 year	0,246	0,000
E. Casualties	E.1 None	0,966	0,992
	E.2 Up to 10 persons	0,034	0,008
	E.3 10–50 persons	0,000	0,000
	E.4 50–200 persons	0,000	0,000
	E.5 More than 200 persons	0,000	0,000
F. Affected citizens	F.1 None	0,000	0,811
	F.2 Up to 10 persons	0,000	0,006
	F.3 10–100 persons	0,021	0,006

(continued)

Table 4.41 (continued)

Parameter	Alternative	Estimate	
		Pipes	Tunnel
	F.4 100–1000 persons	0,427	0,038
	F.5 More than 1000 persons	0,551	0,140
G. Material damage	G.1 Up to 100 MW	0,001	0,464
	G.2 100–1000 MW	0,259	0,512
	G.3 1000–10000 MW	0,431	0,024
	G.4 More than 10000 MW	0,309	0,000
H. Ecological consequences	H.1 No tangible ecological consequences	0,000	0,894
	H.2 Slight, local, short-term worsening of the ecological situation	0,142	0,105
	H.3 Significant long-term worsening of the ecological situation in a large area	0,500	0,001
	H.4 Ecological catastrophe	0,357	0,000

pipes bear very harmful impact for ecology (alternatives H.3—"Significant long-term worsening of the ecological situation in a large area" with weight 0,500, H.4—"Ecological catastrophe" with weight 0,357), denoting much higher ecological risk.

Thus, an underground tunnel for sewage system outperforms underwater pipes under almost all of the criteria, and for some important criteria, this advantage is overwhelming.

The modified morphological analysis method allows also to conduct inference "what-if" analysis, selecting a configuration, or a group of configurations that contain a specific type of threats at the first stage. Respectively, at the second stage, the consequences are shown only for a chosen type of threat, allowing to model and compare different scenarios.

In this study, three scenarios of unfavorable events were taken, determined by the configurations of the MT at the first stage:

Scenario 1 (sabotage through undermining): 1.1—Explosion, 2.1—Anthropogenic with malicious intent, 3.2—Several structural or functional elements, or several areas;
Scenario 2 (technogenic threat): 1.5—Operational damage and/or destruction of object or its parts, 2.3—Technical, technological, 3.2—Several structural or functional elements, or several areas;
Scenario 3 (natural threat): 1.3—Landslides, landfalls, subsidence of soil, 2.3—Technical, technological, 3.3—Whole object.

Also, Scenario 4 was considered—an undefined sabotage, which specifies only the origin of the event—2.1, "Anthropogenic with malicious intent", leaving the exact details undetermined to better understand the multitude of potential military and sabotage threats.

The results of modeling for scenario 1 are shown in Table 4.42.

Convenient comparison of results can be made using diagrams for separate parameters. Scenario 1 results for parameters B, F, G, and H are presented in Figs. 4.14, 4.15, 4.16 and 4.17.

Diagrams allow to compare and evaluate scenarios for underwater pipes and underground tunnels. It is notable that the most disruptive event (explosion) leaves a small chance of full operation for pipes in an underground tunnel (with 0,142 weight), whereas the underwater pipes have zero chance of performing all or a part of functions (Fig. 4.14). Even in case of an explosion, underground tunnel retains high chance of performing a part of functions (weight appr. 0,8). Affecting living conditions of population is the only criterion where the results of underwater pipes and an underground tunnel are relatively close, as disrupting any kind of sewage system will have radical consequences for a large portion of Kyiv population (Fig. 4.15). Material damage for an underground tunnel mostly falls in the alternatives up to 1000 minimum wages (weight 0,87) for repair of casing, hydroisolation, etc., while, for the underwater pipes, an explosion means total destruction with expenses on restoration, and elimination of ecological damage, up to 10000 minimum wages and even more (total weight 0,76—Fig. 4.16). Diagram for parameter H (Ecological consequences) is also very significant. Burst of sewage into the Dnipro River may lead to an ecological catastrophe for the whole river basin. As the diagram in Fig. 4.17 clearly shows, an explosion in an underground tunnel does not impact the ecological situation (weight 0,73), as it lies tens of meters beneath the river bottom, and damage to casing will not impact the situation. Local short-term worsening of ecological situation (weight 0,27) may be caused by exposure of sewage to underground waters, but it does not have a threatening scale. On the other hand, a disruption of underwater pipes causes an ecological catastrophe (weight 0,52) or at least a significant long-term worsening of the ecological situation in a large area (weight 0,47).

Results of morphological modeling corresponding to scenarios 2–4 are shown in Tables 4.43, 4.44 and 4.45 and the respective diagrams (Figs. 4.18, 4.19, 4.20 and 4.21).

Diagrams in Figs. 4.18, 4.19, 4.20 and 4.21 again visibly confirm the advantage of an underground tunnel over underwater pipes, obtained in the modeling results, and this advantage is present in any scenario. Also, the comparison of scenarios shows that intentionally created undesirable events generally cause more harm and lead to higher damage than natural or technogenic events. This methodology may be applied to any other infrastructure objects, laying the foundation of system strategy of urban underground development with the goal of minimizing military, technogenic, and natural threats.

Table 4.42 Scenario 1: 1.1—Explosion, 2.1—Anthropogenic with malicious intent, 3.2—Several structural or functional elements, or several areas

Parameter	Alternative	Estimate	
		Pipes	Tunnel
A. Integrity of the object and its parts	A.1 No damage or negligible damage	0,000	0,009
	A.2 Damage may be undone without interruption of operation	0,000	0,247
	A.3 Damage may be undone with interruption of operation	0,828	0,742
	A.4 Damage is irreversible	0,172	0,001
B. Operational capacity	B.1 Object may perform all of its functions	0,000	0,142
	B.2 Object may perform a portion of its functions	0,004	0,764
	B.3 Object stops functioning	0,996	0,094
C. Potential to transfer functions to other objects	C.1 Object's functions can be transferred without limitations	0,000	0,000
	C.2 Object's functions can be transferred with some limitations	0,000	0,176
	C.3 Object's functions can be transferred with significant limitations	0,403	0,793
	C.4 Object's functions cannot be transferred	0,597	0,031
D. Operation restore time	D.1 Operation restore time is unnecessary	0,000	0,000
	D.2 Operation restore time up to 7 days	0,006	0,447
	D.3 Operation restore time up to 1 month	0,392	0,536
	D.4 Operation restore time up to 1 year	0,471	0,018
	D.5 The object cannot be restored during 1 year	0,131	0,000
E. Casualties	E.1 None	1,000	0,959
	E.2 Up to 10 persons	0,000	0,041
	E.3 10–50 persons	0,000	0,000
	E.4 50–200 persons	0,000	0,000
	E.5 More than 200 persons	0,000	0,000
F. Affected citizens	F.1 None	0,000	0,399
	F.2 Up to 10 persons	0,000	0,000

(continued)

Table 4.42 (continued)

Parameter	Alternative	Estimate	
		Pipes	Tunnel
	F.3 10–100 persons	0,052	0,000
	F.4 100–1000 persons	0,424	0,133
	F.5 More than 1000 persons	0,524	0,468
G. Material damage	G.1 Up to 100 MW	0,000	0,087
	G.2 100–1000 MW	0,239	0,874
	G.3 1000–10000 MW	0,539	0,039
	G.4 More than 10000 MW	0,222	0,000
H. Ecological consequences	H.1 No tangible ecological consequences	0,000	0,729
	H.2 Slight, local, short-term worsening of the ecological situation	0,009	0,270
	H.3 Significant long-term worsening of the ecological situation in a large area	0,470	0,001
	H.4 Ecological catastrophe	0,522	0,000

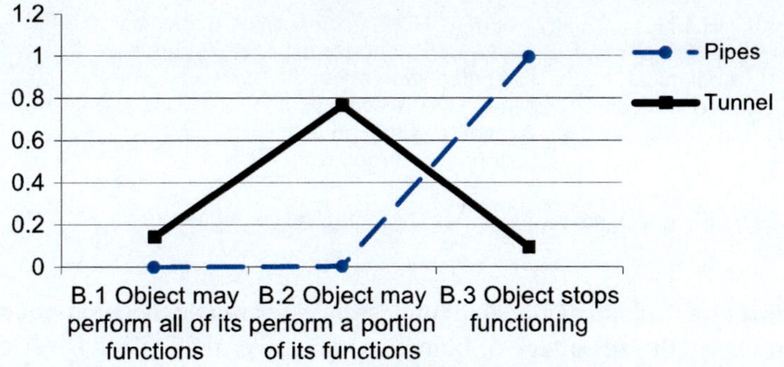

Fig. 4.14 Diagram of weights for parameter B (Operational capacity)

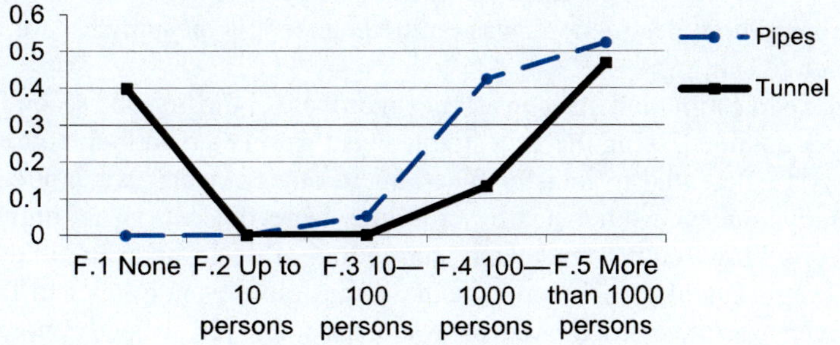

Fig. 4.15 Diagram of weights for parameter F (Affected citizens)

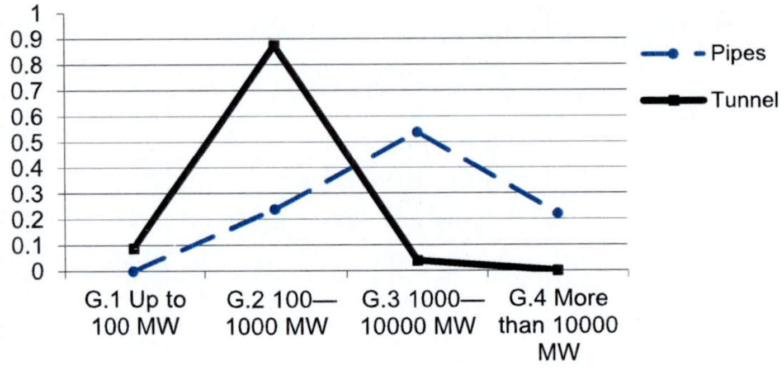

Fig. 4.16 Diagram of weights for parameter G (Material damage)

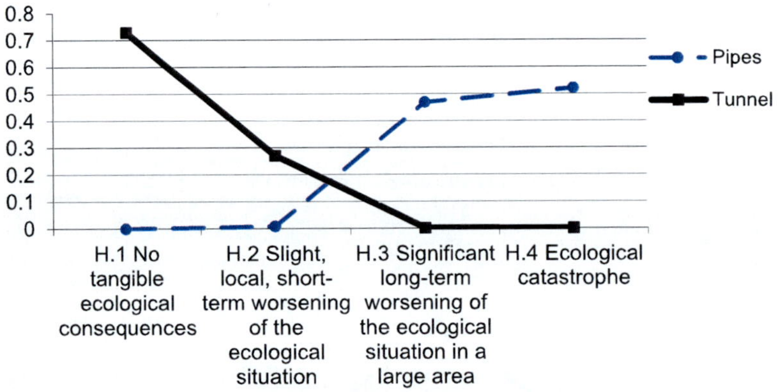

Fig. 4.17 Diagram of weights for parameter H (Ecological consequences)

The same morphological model for undesirable events and their consequences was applied to justify the advantage of tunnel crossing over the Dnipro River compared to the bridge crossing.

The input estimates for the first stage of MMAM, and their calculated values after processing CCM are given in Table 4.46. In this study, the alternative "1.6 Functional interruption without destruction" was omitted, since it is meaningless for bridge or tunnel river crossings.

Using expert estimation, the consistency matrices for the first and second MMAM stages were obtained, tying the undesirable event parameters to their consequences. Parameter "C. Potential to transfer functions to other objects" was not considered in this study, since no other nearby crossings were present, so its only possible alternative is "Impossible" under any configuration of an undesirable event. The result of weight calculation for any potential undesirable event is given in Table 4.47.

The fixed parameter MMAM task was also employed to study more specific undesirable event scenarios. Two scenarios were selected:

Table 4.43 Scenario 2: 1.5—Operational damage and/or destruction of object or its parts, 2.3—Technical, technological, 3.2—Several structural or functional elements, or several areas

Parameter	Alternative	Estimate	
		Pipes	Tunnel
A. Integrity of the object and its parts	A.1 No damage or negligible damage	0,000	0,002
	A.2 Damage may be undone without interruption of operation	0,216	0,383
	A.3 Damage may be undone with interruption of operation	0,649	0,613
	A.4 Damage is irreversible	0,135	0,001
B. Operational capacity	B.1 Object may perform all of its functions	0,000	0,223
	B.2 Object may perform a portion of its functions	0,229	0,628
	B.3 Object stops functioning	0,771	0,149
C. Potential to transfer functions to other objects	C.1 Object's functions can be transferred without limitations	0,016	0,022
	C.2 Object's functions can be transferred with some limitations	0,285	0,573
	C.3 Object's functions can be transferred with significant limitations	0,357	0,382
	C.4 Object's functions cannot be transferred	0,342	0,022
D. Operation restore time	D.1 Operation restore time is unnecessary	0,000	0,000
	D.2 Operation restore time up to 7 days	0,044	0,353
	D.3 Operation restore time up to 1 month	0,312	0,636
	D.4 Operation restore time up to 1 year	0,599	0,010
	D.5 The object cannot be restored during 1 year	0,044	0,000
E. Casualties	E.1 None	1,000	0,995
	E.2 Up to 10 persons	0,000	0,005
	E.3 10–50 persons	0,000	0,000
	E.4 50–200 persons	0,000	0,000
	E.5 More than 200 persons	0,000	0,000
F. Affected citizens	F.1 None	0,000	0,599
	F.2 Up to 10 persons	0,001	0,003

(continued)

Table 4.43 (continued)

Parameter	Alternative	Estimate	
		Pipes	Tunnel
	F.3 10–100 persons	0,055	0,003
	F.4 100–1000 persons	0,447	0,021
	F.5 More than 1000 persons	0,497	0,374
G. Material damage	G.1 Up to 100 MW	0,000	0,459
	G.2 100–1000 MW	0,534	0,510
	G.3 1000–10000 MW	0,427	0,030
	G.4 More than 10000 MW	0,040	0,000
H. Ecological consequences	H.1 No tangible ecological consequences	0,000	0,903
	H.2 Slight, local, short-term worsening of the ecological situation	0,229	0,095
	H.3 Significant long-term worsening of the ecological situation in a large area	0,550	0,001
	H.4 Ecological catastrophe	0,220	0,000

Scenario 1: alternative "1.1 Explosion" is fixed;
Scenario 2: alternative "1.4 Weather cataclysms" is fixed.

The weights of undesirable event consequences are given in Table 4.48 for Scenario 1, and Table 4.49 for Scenario 2.

The results are more visually representative in the form of diagrams (Figs. 4.22, 4.23, 4.24, 4.25, 4.26, 4.27 and 4.28). The line "General" on the graph corresponds to the possibility of any undesirable event configuration.

Like in a previous study, the diagrams are constructed in such a way that shifting the area under the graph to the left means greater favorability of a respective object or scenario.

Diagrams allow to make several conclusions:

- from the reliability regarding the interruption of functioning standpoint (parameters "A. Integrity of the object and its parts", "B. Operational capacity"), the tunnel crossing has a slight advantage over the bridge crossing, but the weather cataclysm scenario specifically stands out, in which the tunnel crossing shows much better results (Figs. 4.22 and 4.23);
- from the restoration in case of undesirable events standpoint, the tunnel crossing also has an advantage over the bridge one, which can be seen in the parameters "D. Operation restore time" (Fig. 4.24) and "G. Material damage" (Fig. 4.27), where the alternative "G.2 100–1000 MW" for a tunnel crossing has an undeniable advantage over others, whereas for the bridge crossing the same situation is for the alternative "G.3 1000–10000 MW", meaning that both in average, and in the

Table 4.44 Scenario 3: 1.3—Landslides, landfalls, subsidence of soil, 2.3—Technical, technological, 3.3—Whole object

Parameter	Alternative	Estimate	
		Pipes	Tunnel
A. Integrity of the object and its parts	A.1 No damage or negligible damage	0,000	0,000
	A.2 Damage may be undone without interruption of operation	0,000	0,000
	A.3 Damage may be undone with interruption of operation	0,832	1,000
	A.4 Damage is irreversible	0,168	0,000
B. Operational capacity	B.1 Object may perform all of its functions	0,000	0,179
	B.2 Object may perform a portion of its functions	0,000	0,536
	B.3 Object stops functioning	1,000	0,286
C. Potential to transfer functions to other objects	C.1 Object's functions can be transferred without limitations	0,005	0,007
	C.2 Object's functions can be transferred with some limitations	0,181	0,579
	C.3 Object's functions can be transferred with significant limitations	0,380	0,402
	C.4 Object's functions cannot be transferred	0,434	0,012
D. Operation restore time	D.1 Operation restore time is unnecessary	0,000	0,021
	D.2 Operation restore time up to 7 days	0,004	0,191
	D.3 Operation restore time up to 1 month	0,179	0,574
	D.4 Operation restore time up to 1 year	0,782	0,213
	D.5 The object cannot be restored during 1 year	0,036	0,000
E. Casualties	E.1 None	0,973	0,993
	E.2 Up to 10 persons	0,027	0,007
	E.3 10–50 persons	0,000	0,000
	E.4 50–200 persons	0,000	0,000
	E.5 More than 200 persons	0,000	0,000
F. Affected citizens	F.1 None	0,000	0,179
	F.2 Up to 10 persons	0,000	0,002

(continued)

Table 4.44 (continued)

Parameter	Alternative	Estimate	
		Pipes	Tunnel
	F.3 10–100 persons	0,023	0,002
	F.4 100–1000 persons	0,459	0,012
	F.5 More than 1000 persons	0,518	0,805
G. Material damage	G.1 Up to 100 MW	0,000	0,194
	G.2 100–1000 MW	0,483	0,719
	G.3 1000–10000 MW	0,427	0,086
	G.4 More than 10000 MW	0,090	0,000
H. Ecological consequences	H.1 No tangible ecological consequences	0,000	0,837
	H.2 Slight, local, short-term worsening of the ecological situation	0,203	0,159
	H.3 Significant long-term worsening of the ecological situation in a large area	0,600	0,004
	H.4 Ecological catastrophe	0,197	0,000

considered specific scenarios the damage from accidents and catastrophes will be approximately ten times higher for bridges than for tunnels;

- the number of casualties (parameter "E. Casualties", Fig. 4.25) is generally equal for both types of objects, with a negligible advantage of tunnels. This parameter is in much greater extent affected by the scenario choice, as in case of explosion the number of casualties predictably increases significantly compared to the general scenario or weather cataclysms;
- the analysis for the parameter "F. Affected citizens" (Fig. 4.26) practically matches the results for the parameter "B. Operational capacity", as the city dwellers are affected in case of partial or complete functioning of the crossing, so the tunnel's advantage here is insignificant, except the weather cataclysm case, where the tunnel crossing demonstrates absolute advantage;
- regarding the parameter "H. Ecological consequences" (Fig. 4.28), the nonzero values are observed only for the alternatives "H.1 No tangible ecological consequences" and "H.2 Slight, local, short-term worsening of the ecological situation", since both types of crossings can't make a serious negative impact on the ecology. However, in this parameter, the advantage of tunnel is also evident.

Therefore, the tunnel crossing concept is more advisable from the standpoints of ecological consequences, resilience to undesirable events of anthropogenic, technogenic, or natural origin, and also technical reliability and longevity.

Table 4.45 Scenario 4: 2.1—Anthropogenic with malicious intent

Parameter	Alternative	Estimate	
		Pipes	Tunnel
A. Integrity of the object and its parts	A.1 No damage or negligible damage	0,000	0,024
	A.2 Damage may be undone without interruption of operation	0,000	0,328
	A.3 Damage may be undone with interruption of operation	0,505	0,647
	A.4 Damage is irreversible	0,495	0,000
B. Operational capacity	B.1 Object may perform all of its functions	0,000	0,152
	B.2 Object may perform a portion of its functions	0,001	0,792
	B.3 Object stops functioning	0,999	0,056
C. Potential to transfer functions to other objects	C.1 Object's functions can be transferred without limitations	0,000	0,003
	C.2 Object's functions can be transferred with some limitations	0,012	0,294
	C.3 Object's functions can be transferred with significant limitations	0,274	0,685
	C.4 Object's functions cannot be transferred	0,714	0,018
D. Operation restore time	D.1 Operation restore time is unnecessary	0,000	0,000
	D.2 Operation restore time up to 7 days	0,002	0,516
	D.3 Operation restore time up to 1 month	0,143	0,474
	D.4 Operation restore time up to 1 year	0,469	0,010
	D.5 The object cannot be restored during 1 year	0,386	0,000
E. Casualties	E.1 None	0,942	0,985
	E.2 Up to 10 persons	0,058	0,015
	E.3 10–50 persons	0,000	0,000
	E.4 50–200 persons	0,000	0,000
	E.5 More than 200 persons	0,000	0,000
F. Affected citizens	F.1 None	0,000	0,712
	F.2 Up to 10 persons	0,000	0,000
	F.3 10–100 persons	0,013	0,000

(continued)

Table 4.45 (continued)

Parameter	Alternative	Estimate	
		Pipes	Tunnel
	F.4 100–1000 persons	0,407	0,100
	F.5 More than 1000 persons	0,580	0,188
G. Material damage	G.1 Up to 100 MW	0,000	0,108
	G.2 100–1000 MW	0,111	0,844
	G.3 1000–10000 MW	0,410	0,047
	G.4 More than 10000 MW	0,479	0,000
H. Ecological consequences	H.1 No tangible ecological consequences	0,000	0,805
	H.2 Slight, local, short-term worsening of the ecological situation	0,011	0,194
	H.3 Significant long-term worsening of the ecological situation in a large area	0,470	0,001
	H.4 Ecological catastrophe	0,519	0,000

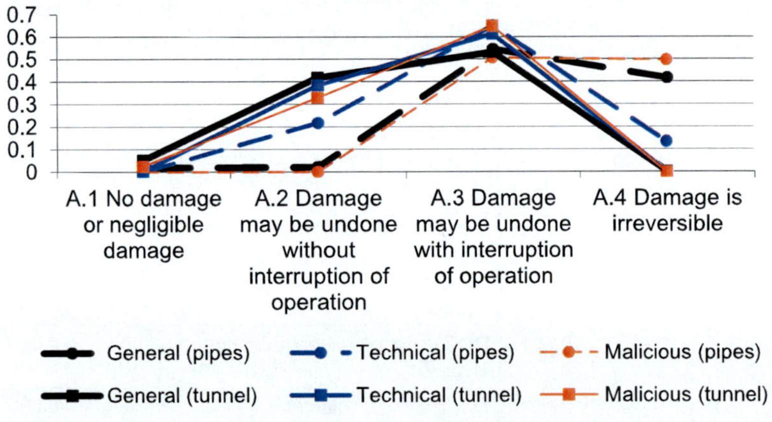

Fig. 4.18 Diagram of weights for parameter A (Integrity of the object and its parts)

4.7 Morphological Table Network for Social Disasters and Catastrophes

In the NATO Program "Science for Peace and Security" supported project NATO.NUKR.SFPP G4877, which was carried out in IASA, a network of morphological tables was used to describe the decision to be made for a social disaster. It can then be used to assess different scenarios of disastrous situations and prepare

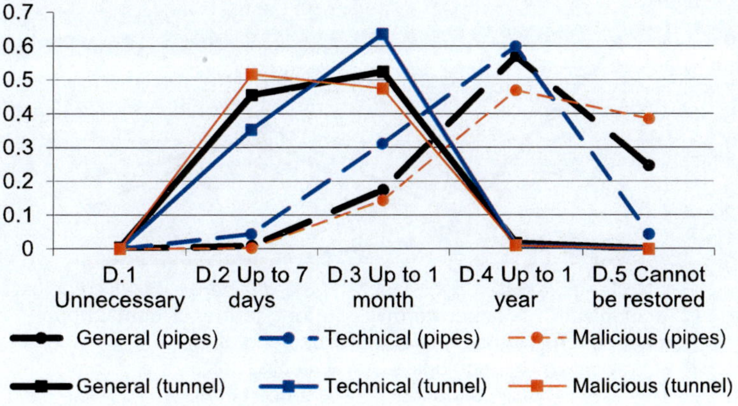

Fig. 4.19 Diagram of weights for parameter D (Operation restore time)

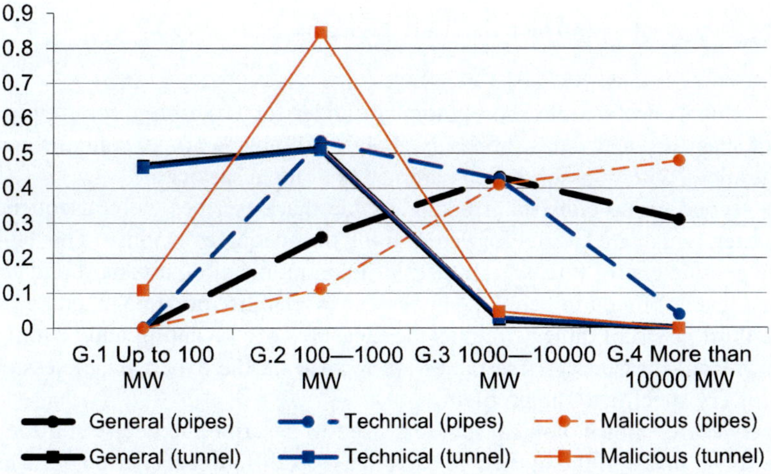

Fig. 4.20 Diagram of weights for parameter G (Material damage)

response strategies for these scenarios. This can be done both for planning multi-disaster management measures and for preparation of strategies against a specific type of disaster.

The developed network consists of three stages represented by morphological tables or groups of tables. The first group consists of a table with common disaster parameters (which are applicable for all disasters), and a table with specific disaster parameters (this table contains parameters which are relevant for the specific type of disasters). This group as a whole describes either a potential multitude of disaster situations with probability estimates, or a specific situation which has to be responded. These tables can be prepared in advance for a number of relevant disaster types.

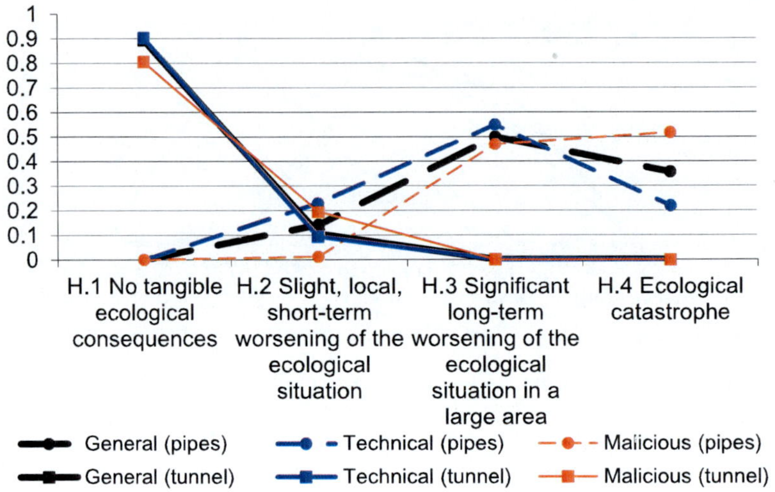

Fig. 4.21 Diagram of weights for parameter H (Ecological consequences)

A fragment of a sample morphological model for describing common parameters of disasters and catastrophes is shown in Table 4.50. It contains possible configurations with non-controlled parameters of the disasters.

The second group consists of a single table that describes the consequences of the disaster, which are crucial for determining the adequate response. This table can include parameters for threats to integrity of communication, transport and control, type and level of threat to people, degree of irreversibility of damage, etc.

The third group of tables contains the measures of preventing and/or mitigating the considered disasters. These tables are formed on the expert panel sessions, as they are very specific to the problem.

A network of morphological tables is used to describe the decision to be made for the social disaster. It can then be used to assess different scenarios of disastrous situations and prepare response strategies for these scenarios. This can be done both for planning multi-disaster management measures and for preparation of strategies against a specific type of disaster.

The network consists of three stages represented by morphological tables or groups of tables. A first group of tables consists of the table with common disaster parameters (which are applicable to all disasters), and a table with specific disaster parameters (this table contains parameters which are relevant for the specific type of disasters). This group as a whole describes either a potential multitude of disaster situations with probability estimates, or a specific situation which has to be responded. These tables can be prepared in advance for a number of relevant disaster types.

The second table describes the consequences of the disaster, which are crucial for determining the adequate response. This table can include parameters for threats to integrity of communication, transport and control, type and level of threat to people, degree of irreversibility of damage, etc. The third group of tables contains

Table 4.46 Input estimates and MMAM calculation result for undesirable event alternatives

Parameter	Alternative	Normalized input values		Values after MMAM calculation	
		Bridge	Tunnel	Bridge	Tunnel
1. Type of undesirable event	1.1 Explosion	0,8	0,65	0,272	0,215
	1.2 Fire	0,65	0,5	0,197	0,164
	1.3 Landslides, landfalls, subsidence of soil	0,35	0,65	0,061	0,230
	1.4 Weather cataclysms	0,8	0,05	0,111	0,006
	1.5 Operational damage and/or destruction of object or its parts	0,8	0,65	0,360	0,385
2. Origin of undesirable event	2.1 Anthropogenic with malicious intent	0,8	0,65	0,382	0,331
	2.2 Anthropogenic without malicious intent	0,8	0,5	0,221	0,031
	2.3 Technical, technological	0,65	0,65	0,179	0,475
	2.4 Natural	0,65	0,05	0,218	0,162
3. Scale of undesirable event's impact	3.1 Separate structural or functional element, or limited area	0,95	0,8	0,522	0,668
	3.2 Several structural or functional elements, or several areas	0,8	0,5	0,271	0,265
	3.3 Whole object	0,65	0,2	0,144	0,062
	3.4 Whole object and neighboring objects	0,5	0,05	0,034	0,003
	3.5 City region and more	0,5	0,05	0,028	0,002

the measures of preventing and/or mitigating the considered disasters. These tables are formed on the expert panel sessions, as they are very specific for the problem.

The connections between the morphological tables are shown in Fig. 4.29.

The preparation step for working with this MAM network in the situation-analytical center (SAC) includes:

1. preparing tables for disasters and consequences in advance by the SAC analysts or during the expert panels;
2. estimation of the values of consistency matrices that connect the disaster parameter tables and the consequences table, the values of the cross-consistency matrix for the tables;
3. partially preparing the table for measures in advance (the final table is generated from the experts' propositions during the specific disaster-related workshop).

Table 4.47 Estimation of undesirable events consequences for all potential types of undesirable events

Parameter	Alternative	Weight	
		Bridge	Tunnel
A. Integrity of the object and its parts	A.1 No damage or negligible damage	0,001	0,003
	A.2 Damage may be undone without interruption of operation	0,083	0,087
	A.3 Damage may be undone with interruption of operation	0,915	0,910
	A.4 Damage is irreversible	0,001	0,000
B. Operational capacity	B.1 Object may perform all of its functions	0,005	0,024
	B.2 Object may perform a portion of its functions	0,252	0,395
	B.3 Object stops functioning	0,743	0,581
D. Operation restore time	D.1 Operation restore time is unnecessary	0,011	0,013
	D.2 Operation restore time up to 7 days	0,476	0,687
	D.3 Operation restore time up to 1 month	0,481	0,296
	D.4 Operation restore time up to 1 year	0,031	0,003
	D.5 The object cannot be restored during 1 year	0,001	0,000
E. Casualties	E.1 None	0,099	0,205
	E.2 Up to 10 persons	0,631	0,651
	E.3 10–50 persons	0,244	0,135
	E.4 50–200 persons	0,026	0,008
	E.5 More than 200 persons	0,001	0,000
F. Affected citizens	F.1 None	0,160	0,175
	F.2 Up to 10 persons	0,007	0,003
	F.3 10–100 persons	0,018	0,003
	F.4 100–1000 persons	0,353	0,359
	F.5 More than 1000 persons	0,462	0,459
G. Material damage	G.1 Up to 100 MW	0,022	0,260
	G.2 100–1000 MW	0,298	0,643
	G.3 1000–10000 MW	0,600	0,096
	G.4 More than 10000 MW	0,080	0,000
H. Ecological consequences	H.1 No tangible ecological consequences	0,220	0,727

(continued)

Table 4.47 (continued)

Parameter	Alternative	Weight	
		Bridge	Tunnel
	H.2 Slight, local, short-term worsening of the ecological situation	0,780	0,273
	H.3 Significant long-term worsening of the ecological situation in a large area	0,000	0,000
	H.4 Ecological catastrophe	0,000	0,000

This preparation allows to quickly launch a complete mathematical MAM procedure, when the exact nature of a problem is defined. This network of morphological tables can be exploited in the situation-analytical center in several different modes, depending on the problem at hand. Three modes stand out:

1. **evaluating preparedness**. The probabilities of a certain disaster parameters are evaluated, and the efficiency of measures is estimated for a hypothetical variety of disaster situations. This shows the most critical problems that may cause the most damage if they happen. The flexibility of this method allows us to make a 'what-if' model, making one or several of the parameters fixed to observe the changes in probability estimates of the other parameters. In this way we can examine the most common properties of the specific type of disaster, or a disaster in a specific region, etc.
2. **monitoring**. This mode is relevant when a threat of a disaster is always present, for example, a social disturbance. The input data for the MAM in this case can be gathered from the available text data sources (media, Internet, etc.) by text analysis tools. The result for the disaster is constantly recalculated, and an early warning may emerge if the estimates for some critical parameters reach certain values. Thus, potentially disastrous situations can be early detected.
3. **reaction**. In this case, the disaster has already happened, and its parameter alternatives are known, so the measures are evaluated according to a known configuration of tables for non-controlled parameters. This mode is useful for a quick reaction and choosing the best methods of mitigation for a disaster.

Assessing decision alternatives using a network of morphological tables

A network of morphological tables was constructed for assessing the decision alternatives regarding a disaster. The "Common parameters" and "Consequences" tables are default for any situation. This allows to reduce the time needed for setup of the method in different circumstances. The network designed for assessing mitigation and prevention of floods is presented in Fig. 4.30.

The "common parameters" describe the parameters of the disaster which are important regardless of the type of the disaster itself: the scale of the disaster, exposition time and expected duration forecast, type of area, etc. The "Common parameters" table that was employed in this model, is shown in Table 4.50.

Table 4.48 Estimation of undesirable events consequences if alternative "1.1 Explosion" is fixed

Parameter	Alternative	Weight	
		Bridge	Tunnel
A. Integrity of the object and its parts	A.1 No damage or negligible damage	0,000	0,003
	A.2 Damage may be undone without interruption of operation	0,026	0,052
	A.3 Damage may be undone with interruption of operation	0,972	0,945
	A.4 Damage is irreversible	0,002	0,000
B. Operational capacity	B.1 Object may perform all of its functions	0,000	0,008
	B.2 Object may perform a portion of its functions	0,161	0,273
	B.3 Object stops functioning	0,839	0,719
D. Operation restore time	D.1 Operation restore time is unnecessary	0,000	0,004
	D.2 Operation restore time up to 7 days	0,443	0,674
	D.3 Operation restore time up to 1 month	0,502	0,316
	D.4 Operation restore time up to 1 year	0,054	0,006
	D.5 The object cannot be restored during 1 year	0,001	0,000
E. Casualties	E.1 None	0,000	0,000
	E.2 Up to 10 persons	0,574	0,655
	E.3 10–50 persons	0,367	0,313
	E.4 50–200 persons	0,056	0,032
	E.5 More than 200 persons	0,002	0,000
F. Affected citizens	F.1 None	0,000	0,064
	F.2 Up to 10 persons	0,011	0,006
	F.3 10–100 persons	0,026	0,006
	F.4 100–1000 persons	0,384	0,355
	F.5 More than 1000 persons	0,579	0,568
G. Material damage	G.1 Up to 100 MW	0,019	0,205
	G.2 100–1000 MW	0,163	0,685
	G.3 1000–10000 MW	0,677	0,110
	G.4 More than 10000 MW	0,142	0,000
H. Ecological consequences	H.1 No tangible ecological consequences	0,199	0,679
	H.2 Slight, local, short-term worsening of the ecological situation	0,801	0,321

(continued)

Table 4.48 (continued)

Parameter	Alternative	Weight	
		Bridge	Tunnel
	H.3 Significant long-term worsening of the ecological situation in a large area	0,000	0,000
	H.4 Ecological catastrophe	0,000	0,000

The table for specific disaster parameters is taken from the repository of morphological tables. This repository is created for types of disasters that are of most concern for the state or the organization that performs this modeling. In our case, a table for floods was taken, presented in Table 4.51.

The "Consequences" table describes the potential outcomes of the disaster situation, and the structure of this table, like "Common parameters", is default for all situations (however, it may be slightly modified for specific cases). It contains the parameters regarding the safety of people (type and level of threat, presence of potential shelter, etc.) and integrity of communications and control (transport, electricity, water, etc.). The estimates for this morphological table are calculated via MMAM procedure in three steps: a) initial estimates are given; b) the influence of two preceding morphological tables is accounted; and c) the cross-influence of parameters in the table itself is accounted. The "Consequences" table in this example is presented in Table 4.52.

The next tier of tables represents the decision alternatives regarding the situation described by a configuration of previous morphological tables. Depending on the method of using morphological analysis, this decision either takes into account a specific disaster situation, or a whole multitude of possible situations to estimate the preparedness for some type of disasters as a whole. These tables are mostly constructed by experts and decision-makers in their respective field of expertise.

The estimation of these tables is made with the same three-step procedure described for the "Consequences" table.

In this case, a table for mitigation measures was considered, shown in Table 4.53.

The proposed application of MMAM on the basis of the introduced concept using networks of morphological tables for studying problems connected with complex, unstructured systems, objects, or situations allows to improve the control and management during disasters and provide decision-making support for social disaster situations. The developed technique may be used in one of three modes: evaluating preparedness, monitoring, and reaction regarding a disaster.

Developed tools allow to reduce the required time, financial and human resources for recovering from the effects of social disasters in conditions under inaccuracy, incompleteness, fuzziness, untimeliness, and contradictoriness of information.

Table 4.49 Estimation of undesirable events consequences if alternative "1.4 Weather cataclysms" is fixed

Parameter	Alternative	Weight	
		Bridge	Tunnel
A. Integrity of the object and its parts	A.1 No damage or negligible damage	0,004	0,107
	A.2 Damage may be undone without interruption of operation	0,047	0,232
	A.3 Damage may be undone with interruption of operation	0,949	0,661
	A.4 Damage is irreversible	0,000	0,000
B. Operational capacity	B.1 Object may perform all of its functions	0,011	0,274
	B.2 Object may perform a portion of its functions	0,333	0,695
	B.3 Object stops functioning	0,656	0,030
D. Operation restore time	D.1 Operation restore time is unnecessary	0,013	0,099
	D.2 Operation restore time up to 7 days	0,601	0,647
	D.3 Operation restore time up to 1 month	0,375	0,254
	D.4 Operation restore time up to 1 year	0,011	0,000
	D.5 The object cannot be restored during 1 year	0,000	0,000
E. Casualties	E.1 None	0,070	0,593
	E.2 Up to 10 persons	0,649	0,399
	E.3 10–50 persons	0,265	0,008
	E.4 50–200 persons	0,015	0,000
	E.5 More than 200 persons	0,000	0,000
F. Affected citizens	F.1 None	0,197	0,924
	F.2 Up to 10 persons	0,006	0,000
	F.3 10–100 persons	0,006	0,000
	F.4 100–1000 persons	0,338	0,032
	F.5 More than 1000 persons	0,453	0,044
G. Material damage	G.1 Up to 100 MW	0,014	0,433
	G.2 100–1000 MW	0,286	0,567
	G.3 1000–10000 MW	0,657	0,000
	G.4 More than 10000 MW	0,043	0,000
H. Ecological consequences	H.1 No tangible ecological consequences	0,082	0,641

(continued)

Table 4.49 (continued)

Parameter	Alternative	Weight	
		Bridge	Tunnel
	H.2 Slight, local, short-term worsening of the ecological situation	0,918	0,359
	H.3 Significant long-term worsening of the ecological situation in a large area	0,000	0,000
	H.4 Ecological catastrophe	0,000	0,000

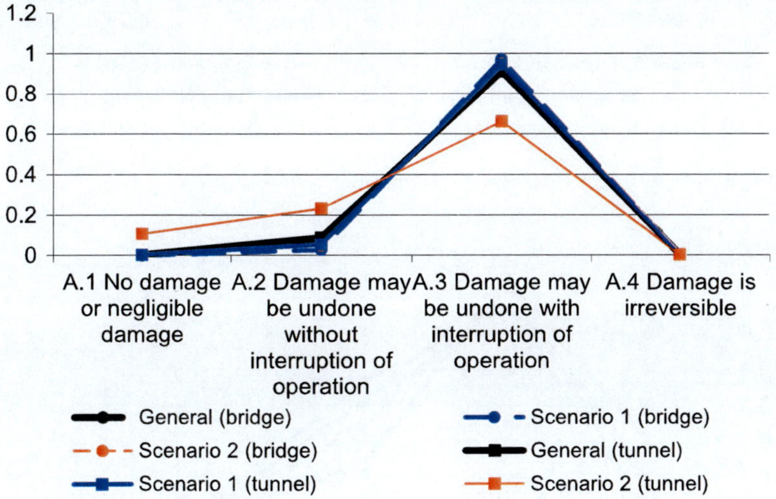

Fig. 4.22 Weight diagram for parameter "A. Integrity of the object and its parts"

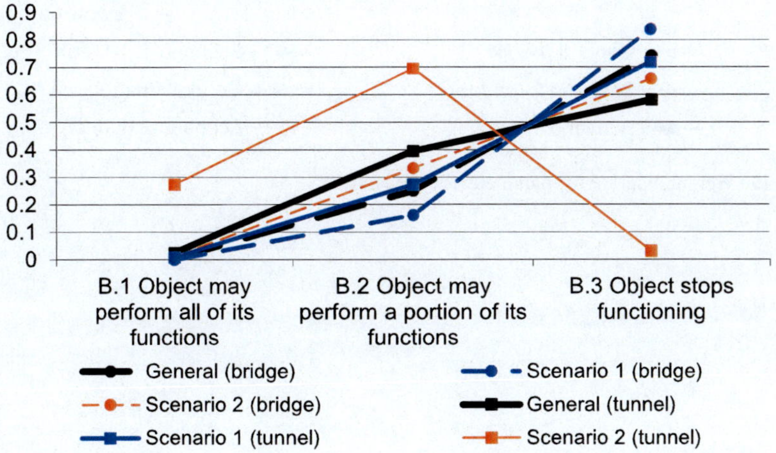

Fig. 4.23 Weight diagram for parameter "B. Operational capacity"

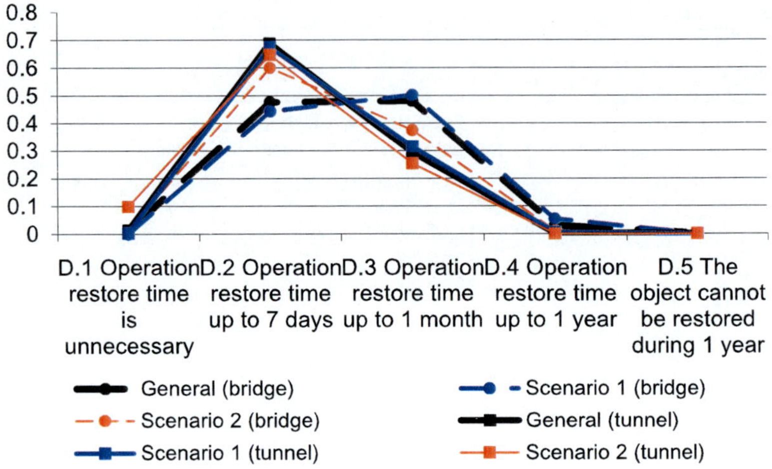

Fig. 4.24 Weight diagram for parameter "D. Operation restore time"

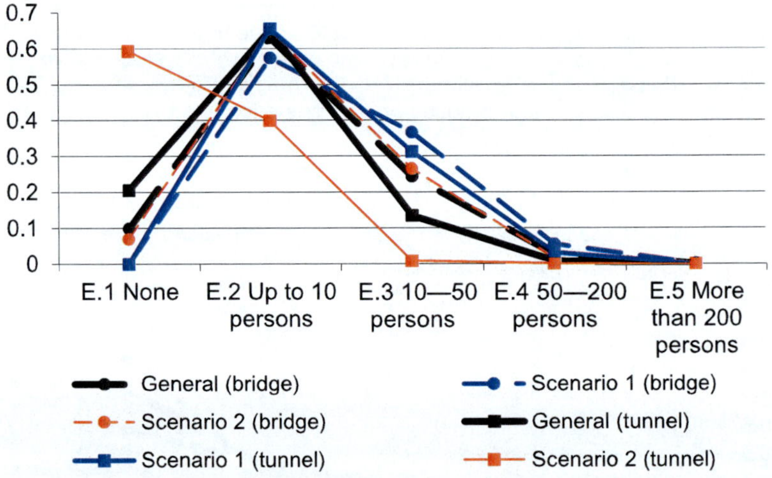

Fig. 4.25 Weight diagram for parameter "E. Casualties"

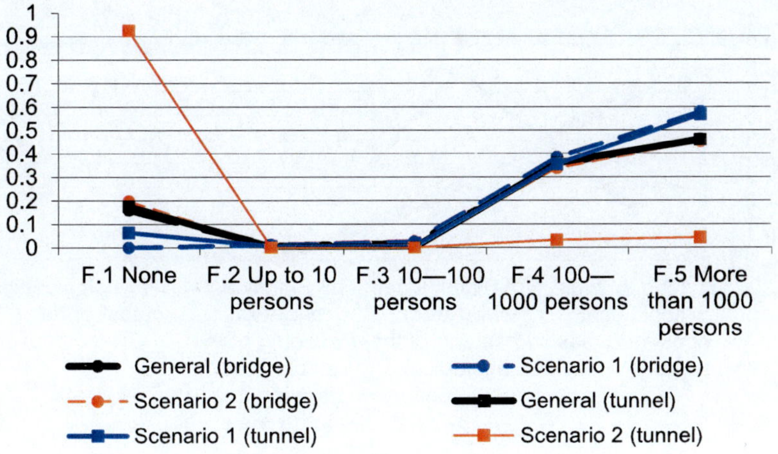

Fig. 4.26 Weight diagram for parameter “F. Affected citizens”

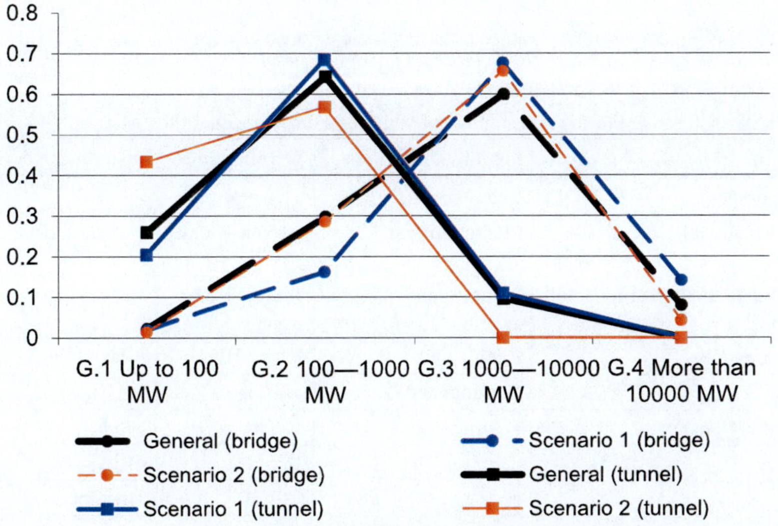

Fig. 4.27 Weight diagram for parameter “G. Material damage”

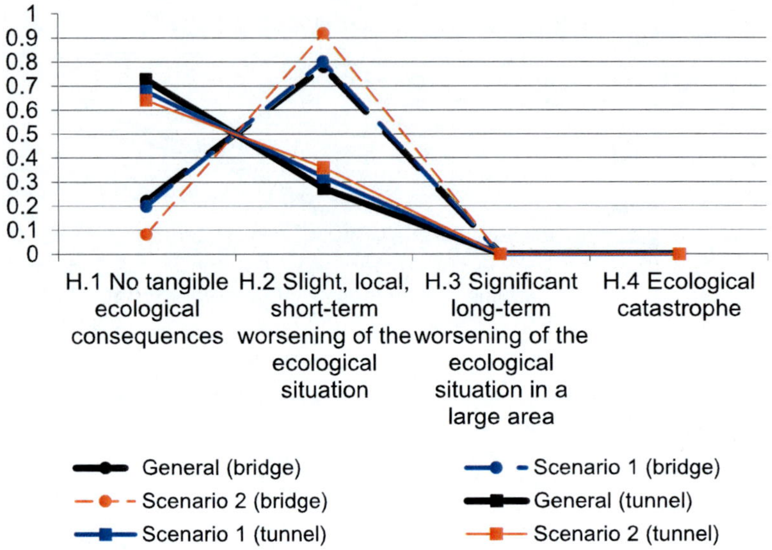

Fig. 4.28 Weight diagram for parameter "H. Ecological consequences"

Table 4.50 A fragment of the morphological table for a disaster situation

Area	Region	Warning time	Disturbance duration
Single object	Heavy urban	>1 week	>1 week
Small area/block/several objects	Medium urban	1 week–3 days	1 week–3 days
Medium area/city district/small town	Light urban/Outskirts	3 days–1 day	3 days–1 day
Large area/city	Rural	1 day–10 h	1 day–10 h
Region	Unpopulated	5–10 h	5–10 h
Several regions		3–5 h	1–5 h
Country		1–3 h	<1 h
		<1 h	Momentary

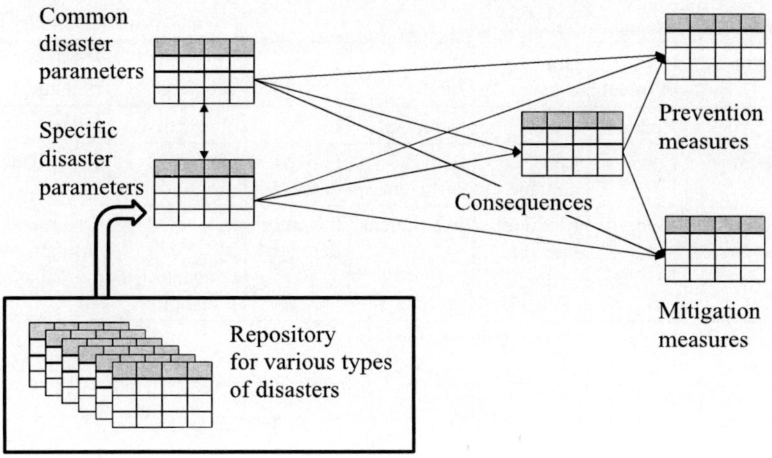

Fig. 4.29 A network of morphological tables for disaster situations

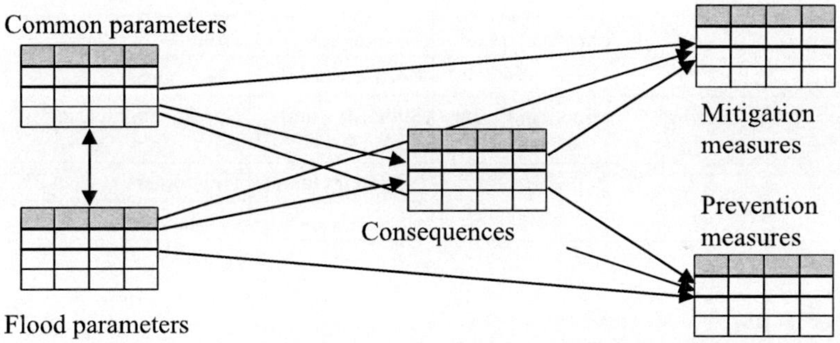

Fig. 4.30 A network of morphological tables for assessing mitigation and prevention of floods

Table 4.51 "Flood parameters" morphological table

Flood type	Class	Main trigger	Cause	Landscape
Slow	Low	Heavy rains	Anthropogenic	Flat
Flash	High	Flood control fail	Natural	Hills
	Extreme	Snow melting		Mountains
	Catastrophic	Hydrological changes		

Table 4.52 "Consequences" morphological table

Type of threat	Degree of threat	Housing	Water supply	Power network	Transport capacity	Civil unrest potential
Direct	None	Intact	Sufficient	Intact	Unrestricted	None
Biological	Low	Low damage	Partly sufficient	Partly damaged	Damaged or rough roads	Disturbances
Chemical	Medium	Medium damage	Insufficient	Severely damaged	Inaccessible by conventional transport	Disorders, insubordination, vandalism
Radioactive	High	Destroyed	Absent/ destroyed			Mass disorders, rebellions
Psychological	Extreme					

Table 4.53 "Mitigation measures" morphological table

Preventing human exposure to disaster	Help	Involved units	Required vehicles, equipment
Evacuation	Medical	Volunteers	None
Temporary shelters	Psychological	Local SES units	Cars, buses
Temporary engineering structures	Food and water	State SES units	Special engineering
		National Guard	Helicopters
			Military

References

Cuia J, Nelson JD (2019) Underground transport: an overview. Tunn Undergr Space Technol 87:122–126

Gilbert P et al (2013) Underground engineering for sustainable urban development. The National Academies Press, Washington

Haiko H (2014) Problems of System Planning of Underground Space in Large Cities. Visnyk NTUU "KPI", Mining series 25:35–40. (In Ukrainian)

Haiko H, Kril T (2015) Typization of geological environment of urbanized territories in mastering of underground space. In: XIV international scientific practical conference "Modern Informational technologies for Managing Ecological Safety, Natural Use, Emergency Measures", Kyiv, pp 173–180. (In Ukrainian)

Haiko H, Bulhakov V, Siveryn M (2016) Construction of a system of road tunnels as a way to solve transport and environmental challenges of a metropolis. Visnyk of NTUU "Kyiv Polytechnic Institute", "Mining" series 30:196–206. (In Ukrainian)

Haiko H, Savchenko I, Matviichuk I (2019) Development of a morphological model for territorial development of underground city space. Naukovyi Visnyk Natsionalnoho Hirnychoho Universytetu 3:92–98

Kelemen Y, Vayda Z (1985) City underground. Stroyizdat, Moscow. (In Russian)

Pankratova N, Savchenko I (2009) Strategy of application of morphological analysis methods in technology foresight process. Naukovi visti NTUU "KPI" 2:35–44 (In Ukrainian)

Pankratova N, Savchenko I, Gayko G, Kravets V (2018) Evaluating perspectives of urban underground construction using modified morphological analysis method. J Autom Inf Sci 50(10):34–46

Pankratova N, Savchenko I, Haiko H, Kravets V (2019) System approach to planning urban underground development. Int J Inf Content Process 6(1):3–17

Radkevych A, Arutunian I, Saikov D (2018) Analysis of concept for forming competitive ability level in contracting enterprises in Ukraine under conditions of dynamic transformations of state construction service market. Bridges Tunnels: Theory Res Pract 14:37–45 (In Ukrainian)

Resin V, Popkov Y (2013) Development of large cities under conditions of transitional economy. In: System approach. Bookhouse "LIBROKOM", Moscow. (In Russian)

Ryabkova E (2014) Design of multistory garages and car parks. TOGU Publisher, Khabarovsk (In Russian)

Samedov A, Kravets V (2011) Constructing Urban Underground Structures. NTUU "KPI", Kyiv. (In Ukrainian)

Vähäaho I (2014) Underground space planning in Helsinki. J Rock Mech Geotech Eng 6:387–398

Chapter 5
Evaluating Ecological Risks of Underground Transport Infrastructure Development Using BOCR

Abstract The priority tasks for human and social safety are considered as minimization problems of ecological and technogenic risks under conditions of constantly growing metropolises. The capabilities of underground structures of taking onto themselves the functions of the most hazardous and risk-bearing surface objects and communications are evaluated. The decision support models for assessing alternative car tunnel tracks in Kyiv are constructed. The BOCR (benefits, opportunities, costs, risks) technique for development of underground traffic infrastructure in Kyiv is developed and implemented. The task of prioritizing models (alternative variants) of the city's underground development is considered, taking into account the conditions of incompleteness, imprecision, and indistinctness of input data. The task incorporates the results of expert estimation for the objects. The priorities for automobile tunnel tracks in Kyiv are calculated by aggregation of a network of parameters, justifying the most rational sequence of their construction, taking into account the criteria of reducing ecological and technogenic risks in urbanized space.

5.1 Introduction

One of the modern characteristic features is the constant growth of metropolises, expansion of urban infrastructure, population growth, and abrupt increase of car numbers, causing significant complications in managing risks in large cities. The most typical threats to the urban environment are the air pollution by car exhaust fumes; dynamic and noise impacts; negative influence of industrial areas, landfills; bursts in sewerage, water, and power supply systems; overloaded treatment plants; traffic accidents and traffic jams; landslides and other harmful geological environment processes; all of these constantly attract attention of researchers and ecological organizations (https://population.un.org/wup/Publications/Files/WUP2018-Report.pdf; https://mepr.gov.ua/files/docs/%D0%9C.%20%D0%9A%D0%98%D0%87%D0%92.pdf; Klymchyk et al. 2016; Korendyaseva 2017). The main cause of cardiovascular and oncological diseases and damage to respiratory organs of city dwellers are the toxic automobile exhaust fumes, and the

N. Pankratova et al., *Modeling the Underground Infrastructure of Urban Environments*, The Urban Book Series, https://doi.org/10.1007/978-3-031-47522-1_5

related noise pollution is the main cause of disorders in auditory and nervous systems; moreover, long-term human exposure to loud noise decreases the work efficiency by up to 60% (Prepotenska 2014; Stolberg 2000). Traffic accidents create the largest risks of injuries and deaths of urban dwellers in peacetime. This is why solving the transport communication problem, particularly the ever-increasing traffic intensity in metropolises, is one of the largest directions for minimizing ecological and technogenic risks in large cities. The global concepts for ecologizing urban environment pay significant attention to the capacity of underground space of taking on the most dangerous and risk-bearing functions of surface objects and communications, as the influence of any underground object on the natural environment is much lower than for its surface analog, and is much better controlled (Golubev 2005; Kartosiya 2015; Gilbert, et al. 2013; Haiko et al. 2018). The planning of underground transport infrastructure should be based on the system methodology and the complex system analysis tool set (Pankratova et al. 2020).

5.2 The BOCR Technique for Estimating Priorities of Model Alternatives

Under conditions of budget limitations for state and city administrations, a task arises for estimating the importance (priority) of the alternative models of urban underground infrastructure development, with further selection of the objects for prior implementation. This problem belongs to weakly structured decision support problems. Their known peculiarity is the lack of the control object's strictly formalized goal of functioning, and as a result, the absence of an optimal solution in its classic understanding; as well as the insufficiency of determined information for obtaining the solution.

The problem of estimating the importance, or priorities, of models (alternative options) for urban underground infrastructure development features uniqueness, incompleteness, imprecision, and indistinctness of input data. Solving it may be based on mathematical methods of analytic hierarchy, where the input information is the combination of quantitative data and expert estimations. In (Nedashkivska 2018; Nedashkivska 2013; Pankratova and Nedashkovskaya 2018, 2016; Pankratova et al. 2021) the constituents of methodology and tool set for decision support are proposed, enabling calculating priorities of alternatives using hierarchical and network models for criteria. Hierarchy decision support and expert estimation models and methods are used to solve the problems of sustainable development of urban territories (Nesticò et al. 2020; Mohammed Ameen and Mourshed 2019), planning metropolises, and underground construction (Doyle 2020; Peng and Peng 2018). The studies (Nesticò et al. 2020; Mohammed Ameen and Mourshed 2019; Doyle 2020; Peng and Peng 2018), though, have some limitations; among them the insufficiently reasoned consistency level; lack of implementation of expert estimation and consistency increase procedures for expert assessments; insufficiently formalized solutions using fuzzy expert assessments; weak analysis of sensitivity and resilience of the obtained solutions. The developed BOCR modification for evaluating decision alternatives (Pankratova and Nedashkovskaya 2018) is free from these limitations.

Let us demonstrate how these methods and tools may be applied to evaluate the priority of models (alternatives) of urban underground infrastructure development in Kyiv city. Using the Kyiv General Development Plan (https://kga.gov.ua/generalnij-plan), the assessments by local experts, and the experience from other countries, a decision support network is constructed, which is the base for the follow-up estimation of priorities for models (alternatives) of urban underground infrastructure development in Kyiv city.

The decision support (DS) network contains the following elements (Pankratova and Nedashkovskaya 2018):

$$S = \{V, L, E, PCM\},$$

- V is a set of the graph nodes: $V = \{C_1, C_2, ..., C_N\}$, where $C_i = \{v_{i1}, v_{i2}, ..., v_{im_i}\}$ is a set of clusters containing separate DS problem elements. For example, a cluster may contain alternative decision elements (e.g., tunnels), and another cluster contains the characteristics and parameters of alternatives used for evaluation, as well as the factors that substantiate a choice of an alternative. These characteristics will further be called the decision criteria.
- L is the set of the graph's oriented arcs. There are two types of relations in a network: the external dependencies between elements of different clusters mark the influence of elements in one cluster on the elements of another cluster; the internal dependencies show the mutual influence between elements of a single cluster.
- E is the set of assessments for graph elements, given by an expert using a scale.
- PCM is the set of inversely symmetrical pair-wise comparison matrices for graph elements, constructed upon the expert assessments from the set E.

Often the DS network also includes the benefits, opportunities, costs, and risks criteria (abbreviated as BOCR), inherent to an implementation of the considered alternatives. Comparing alternative decisions in regard to benefits and opportunities determines the alternative that offers the better benefits or provides more opportunities. Comparing alternative decisions in regard to costs and risks sets the question of which alternative is more costly and risky. In practice, the most common BOCR DS networks contain the components shown in Figs. 5.1 and 5.2.

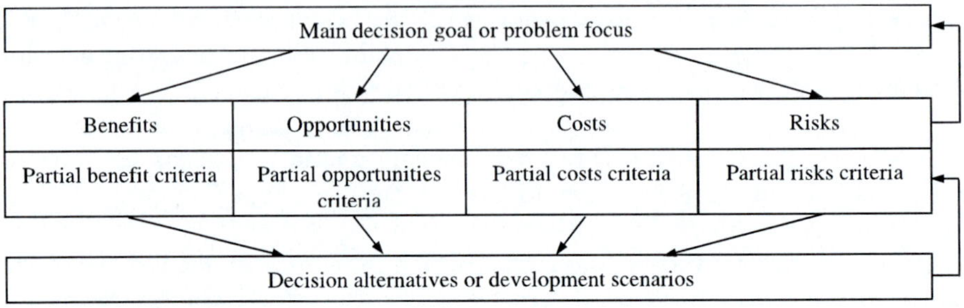

Fig. 5.1 The first version of the BOCR network to evaluate risk-based alternatives

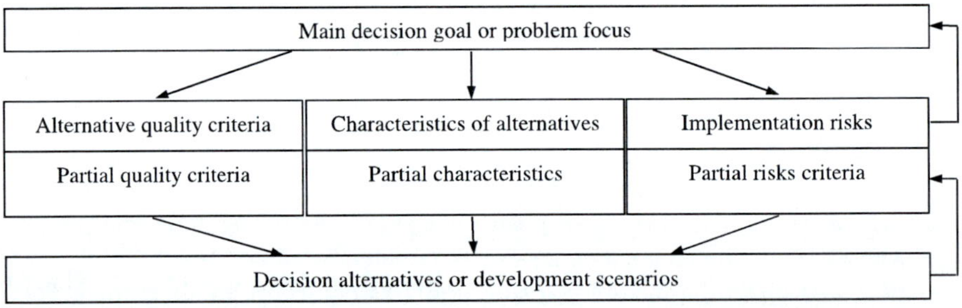

Fig. 5.2 The second version of the BOCR network, taking into account the risks

5.3 The Modified BOCR Method

The features of the method include the calculating of aggregated decision alternative priorities using a DS network, taking into account the risks by pair-wise comparison-based expert estimation; the ability to increase the expert's assessment consistency without the expert's participation; the potential to make a complex sensitivity assessment of solutions obtained on each of the method's steps. As a result, we obtain priorities of alternatives, aggregated by a set of parameters, creating an opportunity to select those that require urgent implementation.

The modified BOCR method steps are described in Pankratova and Nedashkovskaya (2018). The input data of the method consists of the expert assessments, presented in the form of inversely symmetric pair-wise comparison matrices (PCM) $D_{n\times n} = \{(d_{ij})\},\ d_{ij} > 0,\ d_{ji} = 1/d_{ij}$ for the network elements. Let us show the quality analysis for crisp PCMs. In (Nedashkivska 2018), a technique was proposed for increasing the consistency of a crisp PCM D without the expert's participation. We will present it under the assumption that the consistency is estimated using the CI^{tr} index:

1. Set the parameter α value, $0 < \alpha < 1$. Calculate the consistency index CI^{tr} of the PCM D.
2. Iteratively, until the criteria of permissible inconsistency $CI^{tr} < CI^{tr\ porog}$ is met:

2.1. Calculate the weights $w = (w_1, ..., w_n)^T$ using the PCM D.
2.2. Calculate the corrected PCM $D* = (d^*_{ij})$, using the multiplicative or additive correction described below.
2.3. Calculate the CI^{tr} index of the corrected PCM D^*. This matrix becomes the new matrix D.

The multiplicative PCM correction is based on the statement formulated and proven in Nedashkivska (2013). Let $D^*_{n\times n} = (d^*_{ij})$, $d^*_{ij} = (d_{ij})^{\alpha}\left(\frac{w_i}{w_j}\right)^{1-\alpha}$ be the corrected PCM, where $\alpha \in (0,\ 1)$ is the correction parameter. Then $CI^{tr*} \leq CI^{tr}$, where CI^{tr*} is the transitive consistency index of the corrected PCM $D^*_{n\times n}$; CI^{tr} is the consistency index of the initial $D_{n\times n}$; the equality $CI^{tr*} = CI^{tr}$ holds only in case the matrix $D_{n\times n}$ is consistent.

Additive correction is based on a similar statement proven in Nedashkivska (2013), where the corrected PCM $D^*_{n\times n} = (d^*_{ij})$ is calculated as $d^*_{ij} = \alpha d_{ij} + (1-\alpha)\left(\frac{w_i}{w_j}\right)$ for $i < j$, and $d^*_{ij} = \alpha d_{ji} + (1-\alpha)\left(\frac{w_j}{w_i}\right)$ for $i \geq j$.

Let us now consider the estimates formalized with a fuzzy pair-wise comparison matrix (FPCM) $\tilde{D} = \{\tilde{d}_{ij} | i, j = 1, ..., n\}$, where $\tilde{d}_{ij} = (x, \mu_{ij}(x))$ is a fuzzy set representing an advantage intensity of the alternative a_i over a_j, $x \in R$, $\mu_{ij}(x)$ is the value of the membership function for fuzzy advantage of a_i over a_j; $\tilde{d}_{ii} = 1$. The estimation and consistency increase for FPCM is performed using the following method (Pankratova and Nedashkovskaya 2016):

1. Form the defuzzified PCM $D = \{(d_{ij})\} \in R^+_{n\times n}$, where $d_{ij} = Defuz(\tilde{d}_{ij})$, if $\tilde{d}_{ij} \geq 1$, and $d_{ij} = 1/d_{ji}$ otherwise; $i, j = 1, ..., n$.
2. Perform estimation and consistency increase for PCM D, using the described above method for crisp PCMs.

The advantages of the described method, compared to the other existing methods, are the following:

- it provides the opportunity to identify weak FPCM consistency, and assess the permissibility of the FPCM's inconsistency for weight calculation;
- it enables a relatively easy consistency increase for FPCM, the detection of the most inconsistent elements and cycles in FPCMs, based on the methodology for crisp PCMs;
- it doesn't create contradictions in the results obtained from a crisp and a relatable fuzzy PCM, unlike the other methods using the extended binary arithmetic operations; particularly no contradictions occur in a special case $n = 2$;
- it can be applied to FPCM with any type of elements (trapezoid, Gaussian, discrete fuzzy sets, etc.).

Now let us proceed to the method of calculating aggregated priorities of the BOCR network (Pankratova and Nedashkovskaya 2018). A weighted block matrix $WWE = \{(WWE_{ji})\}$ is formed from local priorities of elements in all clusters of the network, where $WWE_{ji} = WE_{ji} \cdot WC_{ji}$, $WE_{ji} \geq 0$ is the matrix of normalized local priorities of elements of cluster C_j regarding the elements of cluster C_i, WC_{ji} is the matrix of normalized weights of cluster C_j regarding the cluster C_i, $j, i = 1, ..., N$. If $I(C_i, C_j) = 0$, then WE_{ji} is a zero matrix. The matrix $WWE > 0$ is stochastic due to the way it's constructed. Using *WWE*, the limit priorities of the network elements are calculated:

- if *WWE* is primitive, then the priorities *w* are the elements of the main eigenvector of the matrix *WWE*;
- if *WWE* is irreducible, imprimitive (cyclic), then the vector of priorities is the average of the matrix WWE^k columns, when $k \to \infty$.

The aggregated priority value for the ith alternative by the model of benefits, opportunities, costs, and risk clusters (Fig. 5.1) is found using one of the rules:

$$\begin{aligned} w_i^{aggr} &= (B \cdot O)/(C \cdot R), \text{ or} \\ w_i^{aggr} &= bB + oO + \frac{c}{C} + \frac{r}{R}, \text{ or} \\ w_i^{aggr} &= bB + oO + (1-c)C + (1-r)R, \end{aligned}$$

where *b*, *c*, *o*, and *r* are the weights for benefits, opportunities, costs, and risks, respectively; *B*, *C*, *O*, and *R* are the local priority values of the ith alternative regarding those qualities.

Similarly, the aggregated priority value for the ith alternative by the model shown in Fig. 5.2 is found using one of the rules:

$$\begin{aligned} w_i^{aggr} &= (Q \cdot H)/R, \text{ or} \\ w_i^{aggr} &= qQ + hH + \frac{r}{R}, \text{ or} \\ w_i^{aggr} &= qQ + hH + (1-r)R, \end{aligned}$$

where q and h are the weights of qualities and characteristics of the decision alternatives, respectively; r is the weight of the alternative implementation risk criterion; Q, H, and R are the local priority values of the ith alternative regarding the defined qualities, characteristics, and risks, respectively.

5.4 Using BOCR Method for Evaluating Tunnel Tracks in Kyiv

Let us now present the application of the BOCR method for evaluating tunnel tracks in Kyiv city. Using the assessments by the local experts, and the experience of other countries, a network decision support model was constructed for evaluating tunnel alternatives in Kyiv; two variants of tunnels were taken as alternatives, namely Tunnel

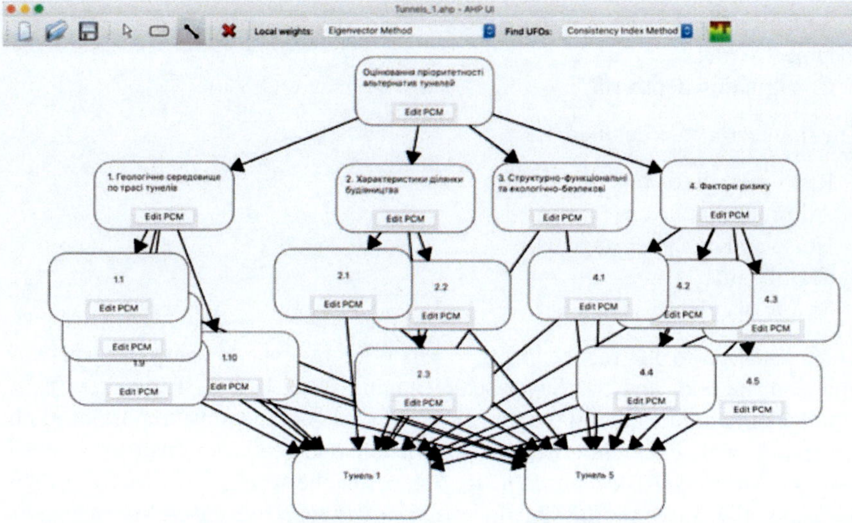

Fig. 5.3 Decision support model for evaluating tunnel alternatives in Kyiv

1 and Tunnel 5 (up to the Dnipro River) from Fig. 4.5, that represent the typical downtown locations.

Evaluating the tunnel alternatives is conducted using the following decision criteria (Fig. 5.3):

1. Geological environment along the tunnel track.
2. Characteristics of the construction site.
3. Structural and functional factors (the tunnel construction expedience), ecological and safety factors.
4. Risk factors.

The parameters that characterize the geological environment along the tunnel track are, in turn, detailed into decision subcriteria:

1.1. Level of dynamic load.
1.2. Static load from surface buildings.
1.3. Static load from soil.
1.4. Influence of existing underground objects.
1.5. Genetic type and lithologic composure of soil.
1.6. Effective soil strength.
1.7. Influence of aquifers and perched groundwater.
1.8. Landscape type and morphometrics.
1.9. Geological engineering processes.
1.10. Geotechnologies of underground construction.

In the decision network, the following construction site characteristics are considered:

2.1. Suitability of the site.
2.2. Object scale.
2.3. Construction depth.

The risk factors are detailed as:

4.1. Risks related to construction site.
4.2. Air pollution.
4.3. Noise and dynamic impacts.
4.4. Traffic jams.
4.5. Traffic accidents.

The evaluation of the model in Fig. 5.3 is done by experts using the pair-wise comparison method, and the Saaty scale ranging from 1 to 9. The criterion "1. Geological environment along the tunnel track" came out the most important of all: it strongly prevails over the criterion "2. Characteristics of the construction site", and weakly over "4. Risk factors" (Fig. 5.4, 5.5). The resulting PCM from these assessments (Fig. 5.6) is permissibly inconsistent, judging from the consistency index of 0,039, less than the threshold value for a PCM of this size. This PCM also has the additional desirable property of weak consistency, and as a consequence, does not contain cycles that cause non-transitive resulting rankings. Figures 5.5 and 5.7 present the calculated local priorities, or weights, of the network elements.

Figure 5.8 also shows a set of assessments with a highlighted throwaway element $a_{21} = 5$. This throwaway element leads to high inconsistency level of assessments, as the PCM inconsistency level spiked to 0,344. The PCM in Fig. 5.8 is contradictory, has conflicting consistency, and cannot be used for further calculations. In Fig. 5.9 the

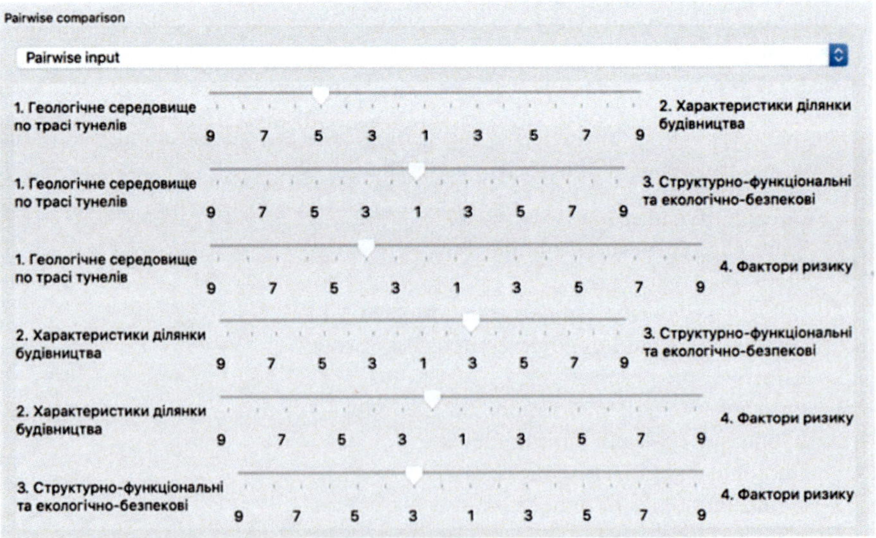

Fig. 5.4 Expert assessments of the decision criteria performed in the adopted scale

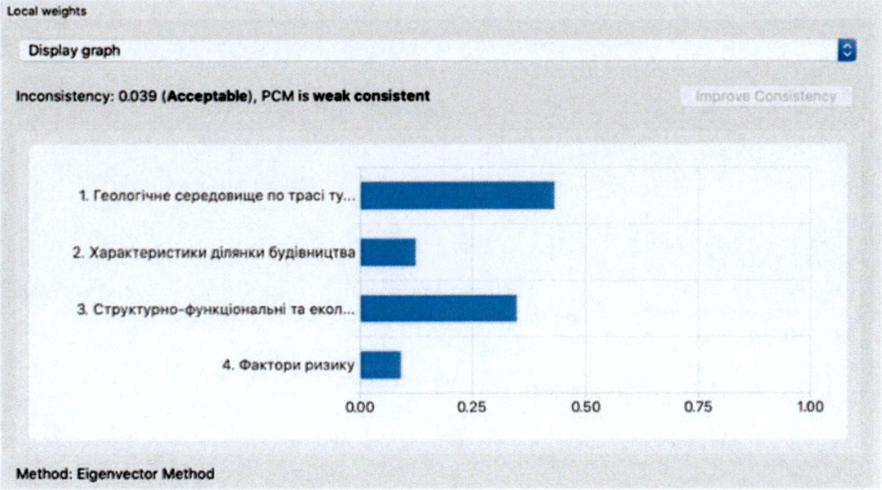

Fig. 5.5 Results of expert evaluation: local weights of decision criteria based on given evaluations

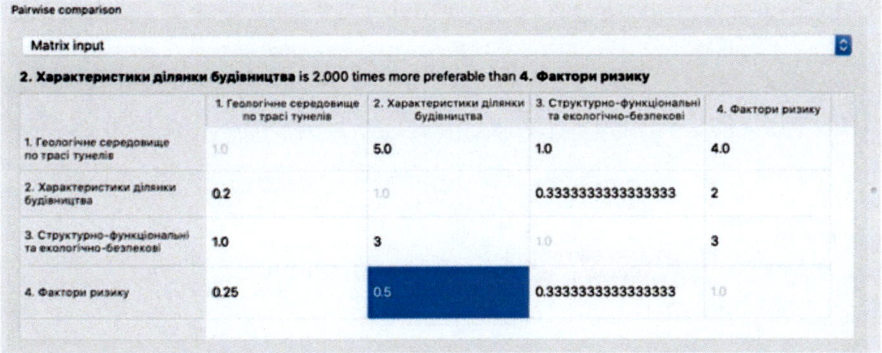
Pairwise comparison

Matrix input

2. Характеристики ділянки будівництва is 2.000 times more preferable than **4. Фактори ризику**

	1. Геологічне середовище по трасі тунелів	2. Характеристики ділянки будівництва	3. Структурно-функціональні та екологічно-безпекові	4. Фактори ризику
1. Геологічне середовище по трасі тунелів	1.0	5.0	1.0	4.0
2. Характеристики ділянки будівництва	0.2	1.0	0.3333333333333333	2
3. Структурно-функціональні та екологічно-безпекові	1.0	3	1.0	3
4. Фактори ризику	0.25	0.5	0.3333333333333333	1.0

Fig. 5.6 BOCR expert evaluation of decision criteria in relation to the main goal

selection of the "Improve Consistency" utility is shown in the application window. As a result, the system automatically finds the most inconsistent expert judgments, or throwaway elements, and proposes more suitable values for them (Fig. 5.10). The final decision of correcting PCM is made by a human—the analyst or the decision-maker. The system also provides manual correction of the assessments, using forms shown in Figs. 5.4 and 5.6. Thus, the decision support system conducts expert data analysis regarding all of the network elements, determines the judgment consistency level, the suitability of their further processing, as well as the tool set for correcting data with or without the expert, toward the goal of increasing their consistency level.

In the given problem, the following priorities for decision criteria were obtained: "C1. Geological environment along the tunnel track" (0,433), "C2. Characteristics

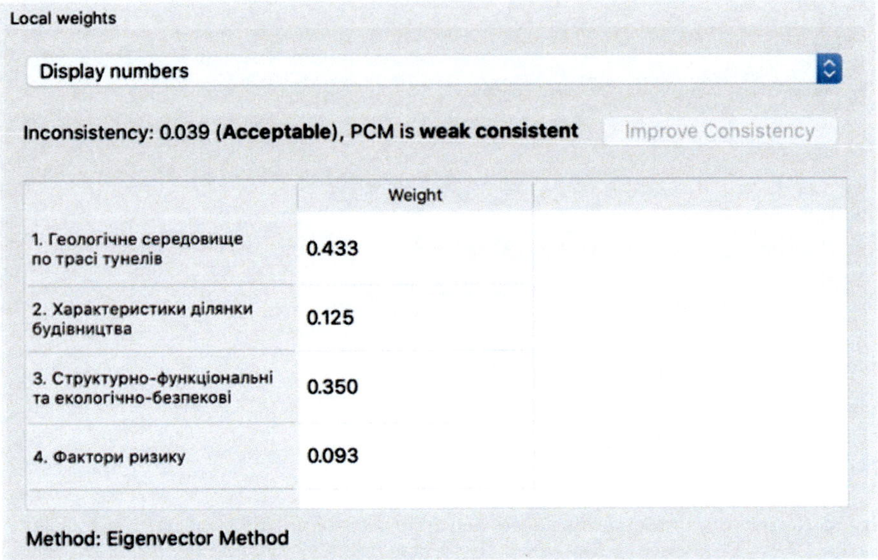

Fig. 5.7 Results of expert evaluation: local weights of decision criteria based on a given BOCR

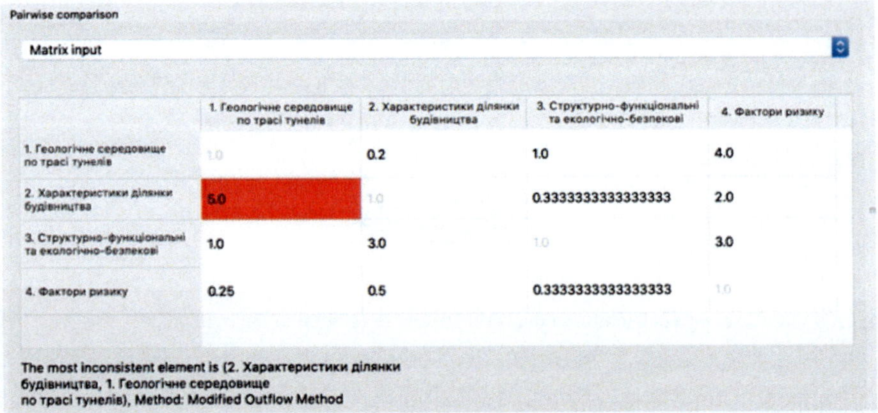

Fig. 5.8 Example of a highly inconsistent BOCR decision criteria

of the construction site" (0,125), "C3. Structural and functional factors, ecological and safety factors" (0,350), and "C4. Risk factors" (0,093), as well as the ranking of criteria based on these values:

$$C1 \succ C3 \succ C2 \succ C4.$$

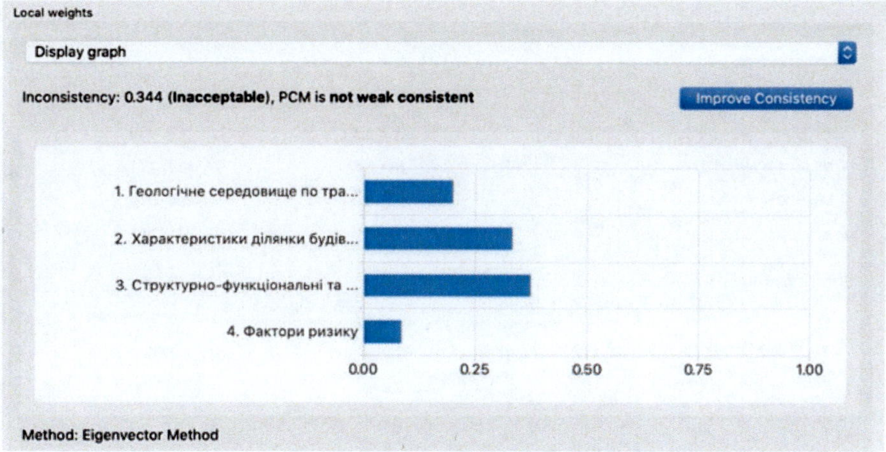

Fig. 5.9 Applying the method of evaluation and increasing consistency to a highly inconsistent BOCR decision criteria

Improved PCM

	1. Геологічне середовище по трасі тунелів	2. Характеристики ділянки будівництва	3. Структурно-функціональні та екологічно-безпекові	4. Фактори ризику
1. Геологічне середовище по трасі тунелів	1.000	3.000	1.000	4.000
2. Характеристики ділянки будівництва	0.333	1.000	0.333	2.000
3. Структурно-функціональні та екологічно-безпекові	1.000	3.000	1.000	3.000
4. Фактори ризику	0.250	0.500	0.333	1.000

Fig. 5.10 Adjusted decision criterion without the participation of the BOCR expert

The sensitivity of the obtained ranking to disturbance in the input data—i.e., expert assessments—can be studied using the "Sensitivity Analysis" utility. The decision support system calculates the intervals and stability indices for keeping consistency (Fig. 5.11), and the intervals and stability indices for keeping the ranking (Fig. 5.12) for each of the PCM elements.

The first type of intervals are those in which the initial assessments may vary for the inconsistency of the whole PCM to remain permissible. The second type of intervals are those in which the initial assessments may vary for the best element or the full ranking to remain immutable. The *stability index* for an element (*i, j*) of a PCM is defined as (Pankratova et al. 2021):

$$\delta_{i,j} = \min((\underline{SInt_{i,j}})^{-1}, \overline{SInt_{ij}}),$$

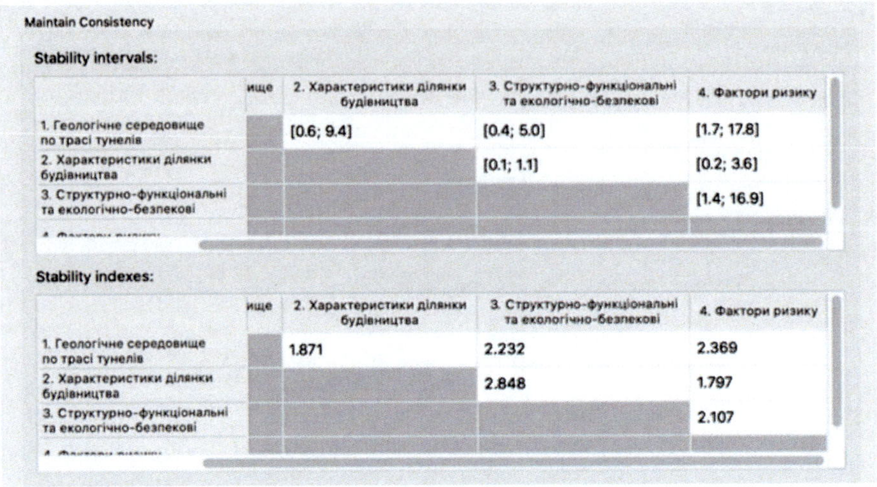
Maintain Consistency

Stability intervals:

	ище	2. Характеристики ділянки будівництва	3. Структурно-функціональні та екологічно-безпекові	4. Фактори ризику
1. Геологічне середовище по трасі тунелів		[0.6; 9.4]	[0.4; 5.0]	[1.7; 17.8]
2. Характеристики ділянки будівництва			[0.1; 1.1]	[0.2; 3.6]
3. Структурно-функціональні та екологічно-безпекові				[1.4; 16.9]

Stability indexes:

	ище	2. Характеристики ділянки будівництва	3. Структурно-функціональні та екологічно-безпекові	4. Фактори ризику
1. Геологічне середовище по трасі тунелів		1.871	2.232	2.369
2. Характеристики ділянки будівництва			2.848	1.797
3. Структурно-функціональні та екологічно-безпекові				2.107

Fig. 5.11 Evaluation of the sensitivity of decision criteria regarding the consistency

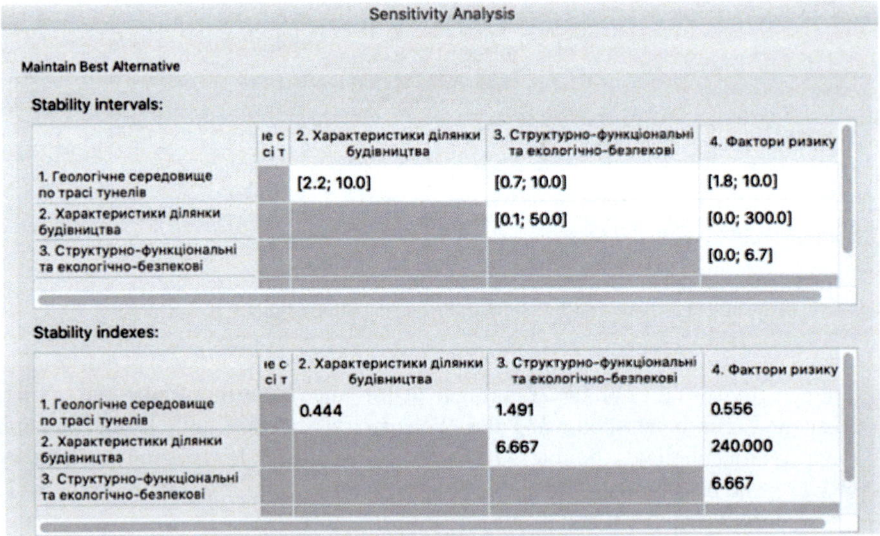
Sensitivity Analysis

Maintain Best Alternative

Stability intervals:

	іе с сі т	2. Характеристики ділянки будівництва	3. Структурно-функціональні та екологічно-безпекові	4. Фактори ризику
1. Геологічне середовище по трасі тунелів		[2.2; 10.0]	[0.7; 10.0]	[1.8; 10.0]
2. Характеристики ділянки будівництва			[0.1; 50.0]	[0.0; 300.0]
3. Структурно-функціональні та екологічно-безпекові				[0.0; 6.7]

Stability indexes:

	іе с сі т	2. Характеристики ділянки будівництва	3. Структурно-функціональні та екологічно-безпекові	4. Фактори ризику
1. Геологічне середовище по трасі тунелів		0.444	1.491	0.556
2. Характеристики ділянки будівництва			6.667	240.000
3. Структурно-функціональні та екологічно-безпекові				6.667

Fig. 5.12 Evaluation of the sensitivity of decision criteria regarding the ranking

where $\underline{SInt_{i,j}}$ and $\overline{SInt_{ij}}$ are the left and right limits of the stability interval, respectively.

So, the most sensitive to consistency level element in a criteria PCM is the element $a_{24} = 2$ in a PCM in Fig. 5.6, as it is characterized by the lowest stability index, equal to 1,797, in a respective matrix in Fig. 5.11. In turn, the most resilient to decision

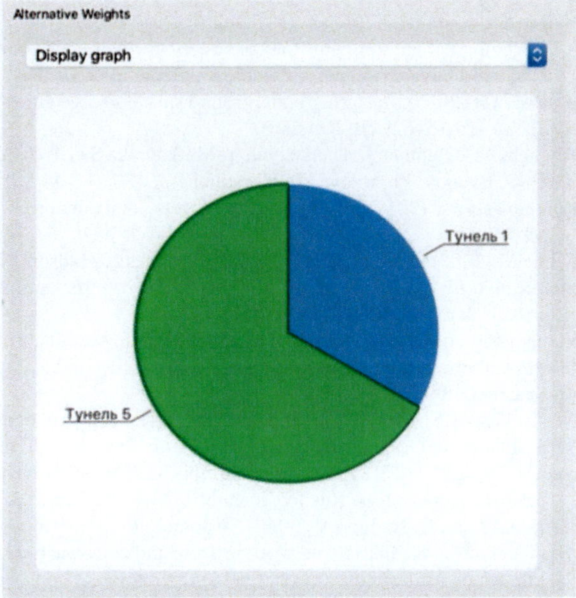

Fig. 5.13 Aggregate priorities of tunnel alternatives based on test expert assessments

criteria PCM consistency change is the element $a_{23} = 0,333$ in the PCM in Fig. 5.6, as it has a maximum corresponding stability index of 2,848 in Fig. 5.11.

The most resilient to changing the obtained ranking $C1 \succ C3 \succ C2 \succ C4$ of the decision criteria is the element a (Shylin 2005; Lehne 2008) of the PCM (Fig. 5.6), with the maximum stability index of 240 in the matrix in Fig. 5.12.

The system calculates all the local weights of the network elements shown in Fig. 5.3, and the results can be seen in the forms similar to Figs. 5.4, 5.5, 5.6 and 5.7.

The result of the calculation for the network shown in Fig. 5.3 is the aggregated priorities of criteria, subcriteria, risks, and finally, the alternative tunnel variants, taking into account all of the stated links. The resulting aggregated priorities for two tunnel alternatives are shown in Fig. 5.13. The diagram visibly shows the priority of the Tunnel 5 construction, as it in larger extent influences the minimization of ecological and technogenic risks.

References

Doyle M (2020) Mapping urban underground potential in Dakar, Senegal: from the analytic hierarchy process to self-organizing maps. Undergr Space 5(3):267–280

Gilbert P et al (2013) Underground engineering for sustainable urban development. The National Academies Press, Washington

Golubev G (2005) Underground urbanism and the city. MIKHiS, Moscow (In Russian)

Haiko H, Matviichuk I, Biletskyi V, Saluga P (2018) Methods of predictive assessment of favorability of geological environment for constructing urban underground objects. V.N. Karazin Kharkiv National University reports, “Geology. Geography. Ecology” series 48:39–51 (In Ukrainian)

Kartosiya B (2015) Mastering underground space of large cities. New Trends Sci Inf Anal Bull (Sci Tech J) 1:615–629 (In Russian)

Klymchyk O, Bahmet A, Dankevych Y, Matkovs’ka S (2016) Ecology of urban systems. Publisher O. O. Evenok, Zhytomyr. (In Ukrainian)

Korendyaseva Y (2017) Environmental aspects of city management. MGUU Moscow Government, Moscow

Lehne M (2008) The complete pyramids. Thames & Hudson, New York

Master plan for the development of Kyiv and its suburbs until 2025 (draft). https://kga.gov.ua/generalnij-plan. Accessed 06 Mar 2023

Mohammed Ameen R, Mourshed M (2019) Urban sustainability assessment framework development: the ranking and weighting of sustainability indicators using analytic hierarchy process. Sustain Cities Soc 44:356–366

Nedashkivska N (2013) A method of consistent pairwise comparisons for decision alternatives evaluation in terms of a qualitative criterion. Syst Res Inf Technol 4:67–79 (In Ukrainian)

Nedashkivs′ka N (2018) A systematic approach to decision support based on hierarchical and network models. Syst Res Inf Technol 1:7–18 (In Ukrainian)

Nesticò A, Elia C, Naddeo V (2020) Sustainability of urban regeneration projects: novel selection model based on analytic network process and zero-one goal programming. Land Use Policy 99

Pankratova N, Nedashkovskaya N (2016) Estimation of consistency of fuzzy pairwise comparison matrices using a defuzzification method. In: Sadovnichiy V, Zgurovsky M (eds) Advances in dynamical systems and control, studies in systems, decision and control, vol 69, pp 375–386

Pankratova N, Nedashkovskaya N (2018) Evaluation of ecology projects for black sea odessa region on basis of a network BOCR criteria model. In: 2018 IEEE first international conference on system analysis & intelligent computing, Kyiv

Pankratova N, Hayko H, Savchenko I (2020) Development of urban underground planning as a system of alternative design configurations. Naukova Dumka Kyiv (In Ukrainian)

Pankratova N, Nedashkovskaya N, Haiko H, Biletskyi V (2021) Assessment of environmental risks of underground transport infrastructure development by BOCR method. V.N. Karazin Kharkiv National University reports, “Geology. Geography. Ecology” series 55:130–143

Peng J, Peng F-L (2018) A GIS-based evaluation method of underground space resources for urban spatial planning: part 1 methodology. Tunn Undergr Space Technol 74:82–95

Prepotenska M (2014) Homo urbanus: the phenomenon of a metropolis human. Seredniak T.K. Publishing, Dnipro (In Ukrainian)

Regional report on the state of the environment of Kyiv for 2017. Kyiv City State Administration. https://mepr.gov.ua/files/docs/%D0%9C.%20%D0%9A%D0%98%D0%87%D0%92.pdf. Accessed 06 Mar 2023

Shylin A (2005) Mastering underground space (genesis and evolution). Mining book, Moscow (In Russian)

Stolberg F (2000) Ecology of the city. Libra, Kyiv. (In Russian)

World Urbanization Prospects 2018: highlights. United Nations, New York, https://population.un.org/wup/Publications/Files/WUP2018-Report.pdf. Accessed 06 Mar 2023

Chapter 6
Strategy for Modeling Complex Urban Underground Environments Based on the Methodologies of Foresight and Cognitive Modeling

Abstract The strategy of urban environments underground construction modeling based on the mathematical support of foresight methodology with the aim of scenarios alternatives creating and cognitive modeling to build scenarios for the development of the desired future and ways of their implementation is proposed. These methodologies are suggested to be used together: the obtained results at the stage of the foresight methodology are used as initial data for cognitive modeling. Using the foresight process at the first stage of modeling allows to identify critical technologies and build alternative scenarios with quantitative characteristics applying expert assessment procedures. For the justified implementation of a particular scenario, cognitive modeling is used, which allows building causal relationships considering a large number of interconnections and interdependencies. The proposed strategy for the development of complex hierarchical systems based on the synthesis of methodologies foresight and cognitive modeling allows to build a science-based strategy to implement priority alternative scenario of complex systems of different natures and offers a unique opportunity within a single software and analytical complex to solve problems of strategic modeling and operational response. The suggested strategy is applied to the study of underground and underwater communications in order to select reasonable scenarios for justification of their creation priority.

6.1 Theoretical Foundation of Foresight and Cognitive Modeling Methodologies

6.1.1 Foresight Methodology of a Complex System

Urban underground modeling is a complex system in many aspects. Firstly, this system consists of many interconnected subsystems and objects. Secondly, the processes occurring in this system during construction and during operation are also complex and, in some cases, poorly predictable, because they are largely associated with various geological processes. The problems that accompany urban underground development can be attributed to poorly structured problems.

N. Pankratova et al., *Modeling the Underground Infrastructure of Urban Environments*, The Urban Book Series, https://doi.org/10.1007/978-3-031-47522-1_6

As a strategy for the innovative development of complex hierarchical systems (CHS) at the level of urban environments, large enterprises, and regions in many countries of the world with a high status, the methodology of foresight has proved itself, which allows us to answer the question "what will happen if ...?" and build alternatives to evidence-based scenarios (Zgurovsky and Pankratova 2007). Given the current trends of the production factor transformation, every large company, industry, or country in the world can not only but must also develop a foresight methodology as a fundamental tool for developing its own policies and strategies in the face of significant changes, new challenges and big risks that the future carries for mankind (Zgurovsky and Pankratova 2005).

The development of the innovation strategy of the CHS belongs to the class of poorly structured tasks in which the goals, structure, and conditions are only partially known and are characterized by a large volume of non-factors: inaccuracy, incompleteness, uncertainty, and vagueness of data describing the object. In contrast to decision-making problems with quantitative values of variables and relations between them, which are solved by methods and means of the operations research theory, econometrics, and other similar methods, specific methods of decision-making support are needed to solve poorly structured problems. The multifactorial, multiparametric, heterogeneous, and poorly structured information of the subject area of the object of study used at different stages of the foresight process leads to difficulties associated with the format for presenting knowledge, constructing questionnaires, processing results, and coordinated management of the foresight process as a whole. Unformalized, heterogeneous, and poorly structured data from the subject area require a single structural description language and a single presentation format (Pankratova and Savastiyanov 2014).

The scenario creation for priority alternative is possible only with the use of a system approach taking into account the totality of the properties and characteristics of the studied systems, as well as the features of the methods and procedures used to create them. Based on a comparison of the characteristics of the qualitative analysis methods, the requirements for their application, and the disadvantages and advantages of each of them, researchers of foresight problems should choose the optimal combination of methods, establish the correct sequence for their use, taking into account the totality of requirements for systems and the features of the tasks to be solved.

The methodological and mathematical support of a systematic approach to solving the problems of developing complex hierarchical systems in the form of a two-stage model based on a combination of foresight and cognitive modeling methodologies is developed and its structural scheme is presented in Fig. 6.1. The involvement of scanning methods, STEEP analysis, brainstorming, SWOT analysis at the first level of the stage allows using expert assessment to identify critical technologies in economic, social, environmental, technical, technological, information and other directions. The basis of this level is the analysis subsystems, which are connected by direct and feedback links to the monitoring system and field tests. The quantitative data obtained after analysis and processing are the initial ones for solving of foresight tasks. At the second level, using the qualitative methods of (morphological analysis,

Delphi, hierarchy analysis (AHP) and its modification, cross-analysis, etc.), the tasks of assessing the behavior of CHS and preparing for decision-making in the form of alternatives to scenarios are solved (Pankratova and Savchenko 2015; Pankratova and Malafieieva 2017; Pankratova and Nedashkivska 2010; Gorelova and Pankratova 2015).

The selection of critical technologies and the construction of rational alternatives of scenarios for the development of strategically important underground urban environments is expedient to be performed on the basis of a collection of foresight.

activities. For this goal, in the process of creating alternatives of scenarios for solving foresight problems, it becomes necessary to involve expert assessment methods, among which are the most commonly used methods of SWOT analysis (Mikhnenko 2015; Alptekin 2013), analytic hierarchy (Pankratova and Nedashkivska 2010), Delphi methods (Pankratova and Malafieieva 2017), cross-impact analysis (Weimer-Jehle 2006), and morphological analysis (Pankratova and Savchenko 2015). For the construction of scenarios that correspond to selected alternatives, cognitive modeling is involved (Gorelova and Pankratova 2015), which makes it possible to obtain a valid scenario for decision-making based on the proposed mathematical apparatus with practical accuracy.

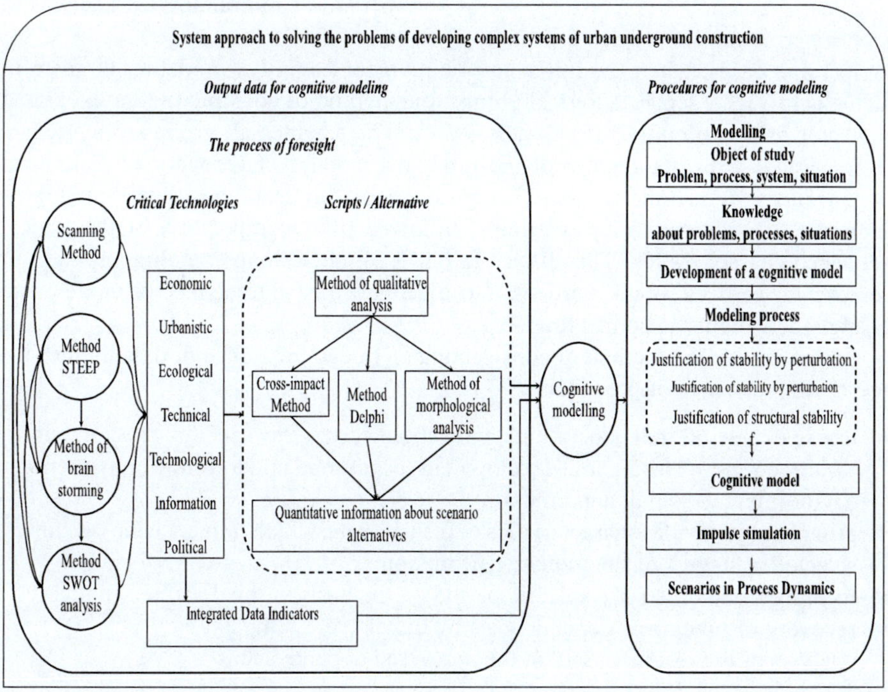

Fig. 6.1 Structural scheme of system approach to solving the problems of developing complex systems of urban underground construction

In these investigations to identify critical technologies, the SWOT analysis and morphological analysis methods are used. For the purpose of ranking the obtained critical technologies and identifying the most topical ones, the TOPSIS (Technique for Order Preference by Similarity to an Ideal Solution) method is applied (Alptekin 2013). The method TOPSIS of multicriterial analysis (ranking) of alternatives in addition to estimating the distance from the considered alternative to the ideal solution allows to take into account the distance to the worst solution. The trade-off in choosing the best alternative is based on the fact that the chosen solution must be at the same time as close to the ideal as possible and most remote from the worst solution. The obtained rating makes it possible to take into account the weight characteristics of critical technologies that are the vertices of the cognitive map when constructing a cognitive model. According to the VIKOR (VIseKriterijumska Optimizacija I Kompromisno Resenje) method, a compromise solution to the problem should be an alternative that is closest to the ideal solution. Moreover, to assess the degree of the alternative proximity to the ideal solution, a multicriteria measure is used (Mardani et al. 2016). As soon as the critical technologies are identified is crossing to the system approach second level, using the qualitative methods for creatiing alternatives of CHS scenarios (Pankratova and Savchenko 2015; Pankratova and Malafieieva 2017; Pankratova and Nedashkivska 2010; Gorelova and Pankratova 2015). For many practical problems, it is advisable to use the method of morphological analysis to identify critical technologies.

In some cases, when the initial information for cognitive modeling is given in statistical form as separate logical groups, the method of constructing an integrated indicator data is proposed (Pankratov 2014). This enables all groups to aggregate in integrated indicator data used the proposed method of recovery of functional dependences for discrete specified samples or carry out decomposition of integrated indicator to individual subject groups, followed by decomposition of the logical sequence characteristics. This allows to build cognitive maps, intelligently add or remove its vertices, break vertices into a sequence of interconnected nodes, and change the weights of connecting arcs.

Formalization the method of constructing an integrated indicator data implies the use of this sequence of procedures:

- the selection of indicator which will characterize the specific area of one of the directions of sustainable development (economic, underground constructions, environmental, social, and others);
- grouping by specific characteristics of the data sets, which influence the dynamics, selected at stage 1 of the indicator formation;
- forming a database for a specific period on the basis of the discrete samples;
- recovery of functional dependencies by the discrete samples;
- analysis of the results based on the recovered dependence.

Based on the expert procedure analysis of innovation activities, the following principles can be used. Instead of the potential realization principle, the possible realization principle is suggested (Zgurovsky and Pankratova 2007). For certain innovation objects (scientific ideas or technical solutions, projects of industrial products, or manufacturing technologies), the initial expert estimation results cannot guarantee their practical feasibility or prove the impossibility of being realized. This principle postulates that for the listed innovation objects, based on the expert estimation results of the presented information, a reliable estimation cannot be obtained a priori that would allow one, for the object under investigation, being grounded and valid, to exclude the possibility of being unrealizable. The estimation for an innovation project retains the uncertainty of the conclusion about the possibility of realization, until, theoretically or experimentally, the possibility of technical or technological realization is proved for the product.

Instead of the truth invariance principle, which postulates the invariability of the theoretical or technical statement, judgement, conclusion, or opinion about the object for a comparatively long time, a completely different principle is needed. Such a necessity is brought about by the previous principle and the innovation activity practice, as expert estimations under conditions of conceptual uncertainty cannot stay the same for a long time. In scientific research and experimental design processes, not only new knowledge about a product under development is accumulated but also the conception about the product's characteristics, use, and application areas may change; new inventions, technical solutions, and other know-how may emerge. Thus, the new principle must reflect the probabilistic characteristics of invariance in time of the initial estimation results of an innovation object, and that is why we shall call it the probabilistic invariance principle. Probabilistic invariance principle. Initial expert estimation results of certain innovative ideas or technical solutions, industrial products, or manufacturing technologies are probabilistic and do not guarantee that they will be saved in time. This principle postulates that initial expert estimation results obtained under conditions of conceptual uncertainty as positive or negative findings, proposals, or recommendations are not invariant and may substantially change, be confirmed, or be disproved as time passes. Therefore, we do not exclude in a certain time frame both safekeeping of truth expert statements, opinions, or findings, and the possibility of disproving them.

The following definitions of latent indices of innovation products are proposed. Practical necessity: the presence of a relatively high market need in innovation products that are proposed in the investigated project or that have a certain demand and sale in domestic and foreign markets. Technology possibility: the presence or possibility to develop materials and components, equipment, and technologies for serial production of innovation products. Economy expediency: the presence of real conditions and the proved prospects of the demand and selling market to obtain an acceptable level of technical and economic effectiveness of innovation products.

At the second stage of system approach for the construction of scenarios corresponding to selected alternatives, cognitive modeling is involved (Gorelova and Pankratova 2015; Roberts 1978), which makes it possible to obtain a valid scenario for decision-making.

6.1.2 Methodology of Cognitive Modeling of Complex Systems

The methodology of cognitive modeling of complex systems was developed while taking into account the properties of socio-economic, socio-technical, ecological, political, and other complex systems and those patterns to which these systems are subject (Gorelova and Pankratova 2015). From these positions, the developed cognitive methodology implements an interdisciplinary approach; it is also a system that unites models, approaches, and methods for solving problems in many areas. The methodology is based on the metamodel of the study, it is a collection of cognitive models (cognitive map, weighted oriented graph, functional graphs, etc.) and methods that include an analysis of the stability of paths and cycles, connectivity, complexity, sensitivity of the system, scenario analysis to foresight the possible development of situations in the system, decision-making, etc. (it includes a software system for cognitive modeling). The toolkit of cognitive modeling has been tested in studies on many complex systems (Gorelova and Pankratova 2018; Pankratova et al. 2019b; Ginis et al. 2016; Abramova and Avdeeva 2008; Avdeeva and Kovriga 2018; Kovriga and Maksimov 2001).

Cognitive modeling is carried out in the following steps.

Pre-project step: collection of information, determination of the nature and development trends of the system systematization of theoretical conceptual positions, analysis of the main properties and characteristics of the system, formulation and clarification of the research objectives.

First Step. Cognitive model development. Designing of the cognitive model *G*; definition of vertices, relations between them, weights, functional dependencies (building a cognitive model of a complex system in the form of a cognitive map or building a cognitive model of a complex system in the form of a parametric functional).

Second Step. Analysis of the properties of the system on cognitive model *G* (determination of the properties of the stability of the model *G* to disturbances, definition and analysis of cycles and structural stability of the model *G*, definition and analysis of paths on *G*, and simplex analysis of the model *G*).

Third Step. Scenario analysis, impulse modeling.

The completion of cognitive modeling is development and evaluation of management decisions to improve the development processes of the system under study.

Let us consider these steps in more detail.

The first step. Development of the cognitive model. In the process of developing a cognitive model, there is a cognitive structuring of the subject area. This is the identification of objects (elements, concepts, essences) of the target system, desirable and undesirable states of the system, the most significant (basic) management factors, environmental factors that affect the transition of the system from state to state, and the establishment on the qualitative/quantitative level of connections (mutual influences) between objects. This is a cyclic process. It begins with the development of simpler mathematical forms of a complex system in the form of a cognitive map G, for which the results of the foresight phase are used as input data. This process can end with more complex forms, for example, in the form of parametric vector functional graphs. For clarity of the further presentation, we give several well-known formulas (Gorelova and Pankratova 2018; Pankratova et al. 2019b; Ginis et al. 2016; Abramova and Avdeeva 2008; Avdeeva and Kovriga 2018; Kovriga and Maksimov 2001; Kulba et al. 2002; Maksimov 2001).

Parametric functional vector graph is a tuple

$$G_F = < G, X, F, q >, \tag{6.1}$$

where $G = <V, E>$ is the cognitive map, where V is the set of vertices (objects, concepts), vertices $Vi \in V, i = 1, 2, \ldots k$ are elements of the investigated system; E is the set of arcs, $e_{ij} \in E$, $i, j = 1, 2, \ldots, n$ reflects the interconnection between vertices V_i and V_j; X is the set of vertices parameters $X = \{x(V_i)\}$, that is, each vertex is associated with a vector of independent parameters $x(V_i)$; $X{:}V \to \theta$, θ is the space of the parameters of vertices, the set of real numbers $F = F(X, E) = F(x_i, x_j, e_{ij})$ is the functional of the transformation of arcs, which assigns to each arc either a sign («+», «−»), then this is a sign digraph, or the weighting factor ω_{ij}, then it is a weighted sign digraph, or a function $F(x_i, x_j, e_{ij}) = f_{ij}$, then this is a functional graph. In the ongoing studies of cognitive modeling for complex systems for constructing models of type (6.1), the idea of composing cognitive maps with known models of system dynamics was realized (Roberts 1978; Kulba et al. 2002; Maksimov 2001; Atkin and Casti 1977).

The cognitive map G corresponds to the square matrix of relations A_G in the following form:

$$A_G = \{a_{ij}\} = \begin{cases} 1, & if\ V_i\ is\, connected\ with\ V_j \\ 0, & otherwise \end{cases} \tag{6.2}$$

The ratio a_{ij} can take the value "+ 1" or "–1". The relation between variables (interaction of factors) is a quantitative or qualitative description of the effect of changes in one variable on others at the corresponding vertexes.

Vector functional graph has the form:

$$\Psi = \langle G, X, F(X, E), \theta \rangle \tag{6.3}$$

where G is a cognitive map; X is the set of vertex parameters, θ is the space of vertex parameters, and F (X, E) is the arc transformation functional. If

$$F(X, E) = F(x_i, x_j, e_{ij}) = \begin{cases} +\omega_{ij}, \textit{ if rising/falling } X_i \\ \textit{entails rising/falling } Xj, \\ -\omega_{ij}, \textit{ if rising/falling } X_i \\ \textit{entails falling/rising } Xj \end{cases} \tag{6.4}$$

then there is a weighted sign digraph, in which ω_{ij} is the weight coefficient.

In the process of studying complex systems various hierarchical cognitive models (6.1) can be developed, as well as models of cognitive models' interaction (6.2) (Gorelova and Pankratova 2018; Pankratova et al. 2019b; Ginis et al. 2016; Abramova and Avdeeva 2008; Avdeeva and Kovriga 2018; Kovriga and Maksimov 2001).

$$I_G = \langle G_k, G_{k+1}, E_k \rangle, k \geq 2 \tag{6.5}$$

where I_G is the hierarchical cognitive map, G_k and G_{k+1} are the cognitive maps of k and $k + 1$ levels, some vertices of which are connected by arcs e_k. If the arc transformation functionals are defined, then a hierarchical cognitive model takes place.

Note that the hierarchical cognitive map model can represent the levels of the management hierarchy of the investigated system, and the lower levels can be a cognitive map that "unfolds" the cognitive top-level map.

The second step. Analysis of the system properties of the cognitive model G. The following are investigated: the stability to disturbances, structural stability, paths, cycles, complexity, connectivity (q-connectedness analysis), sensitivity, etc. (Roberts 1978). The results of the analysis are compared with the available information on complex system.

The stability of the graph by the perturbation and value is based on the concept of the process of propagation of perturbations of a graph. Determine the value at the vertex V_i at the moment of time t through $V_i(t)$, $i \in [1, n]$, $t = 0, 1, \ldots$. Suppose that the value $V_i(t + 1)$, depends on $v_i(t)$, and on the vertices adjacent to V_i. Thus, if a vertex V_j is adjacent to V_i and if $p_j(t)$ represents the change in V_j at the moment of time t, it should be considered that the impact of this change on V_i at the moment of time $t + 1$ will be described by the function $f(V_i, Vj)P_j(t)$, where through $f(V_i, Vj)$ the weight function of connection between the vertices V_j and V_i, is denoted (Roberts 1978). Thus, we have the following rule of perturbation propagation:

$$\begin{gathered} V_i(t+1) = V_i(t) + \sum_{j=1}^{N} f(V_j, V_i) \cdot P_j(t) \; \forall i = \overline{1, n}, \\ P_j(t+1) = V_j(t+1) - V_j(t). \end{gathered} \tag{6.6}$$

The vertex is called stable by perturbation, if the sequence $\{|P_j(t)|\}_{t=1}^{\infty}$ is limited. The vertex is called stable by value if the sequence $\{|V_j(t)|\}_{t=1}^{\infty}$ is limited. The graph is stable by perturbation (value), if all its vertices are stable. Such a result: from the stability by value should be the stability by the perturbation. Thus, the stability by value is reduced to a limited matrix series, and the stability by the perturbation—to limited matrix sequence (Roberts 1978).

Thus, the stability by value is reduced to a limited matrix series $\sum_{t=0}^{\infty} A^t$, and the stability by the perturbation—to limited matrix sequence $M_t = \{A^t\}_{t=1}^{\infty}$ (Weimer-Jehle 2006).

Take the following stability criterion by the perturbation and value.

Criterion 1. The system in the form of signed weighted directed graph G with the adjacency matrix A is stable by perturbation IFF the spectral radius of the adjacency matrix is $p(A) = \max_i |\lambda_i| \le 1$, where $\{\lambda_i\}_{i=1}^{M}$—the eigenvalues A, and is the basis of eigenvectors.

Let us consider the existing possibilities of mathematical analysis for the stability description of the system (Zgurovsky and Pankratov 2014). We represent the expression (6.6) in matrix form:

$$\begin{aligned} V(t+1) &= V(t) + A \cdot P(t), \\ P(t+1) &= V(t+1) - V(t), \end{aligned} \tag{6.7}$$

where A is the adjacency matrix of the graph, $V(t)$ is the vector of values at the vertices $V_1, V_2, \ldots, V_n$ at the time instant t, $P(t)$ is the vector of actions at the vertices $V_1, V_2, \ldots, V_n$ at the instant t. Performing sequential transformations in (6.7), we have

$$\begin{gathered} V(1) = V(0) + A \cdot P(0),\ V(2) = V(1) + A \cdot P(1) = V(0) + A \cdot P(0) + A \cdot P(1), \\ P(1) = V(1) - V(0) = A \cdot P(0),\ P(2) = V(2) - V(1) = A \cdot P(1) = A^2 \cdot P(0), \\ V(t+1) = V(0) + (A + A^2 + A^3 + \cdots + A^{t+1})P(0) = \\ = V(0) + (I + A + A^2 + A^3 + \cdots + A^t)P(1); \end{gathered} \tag{6.8}$$

$$P(t+1) = A^{t+1} \cdot P(0). \tag{6.9}$$

Thus, the stability in value was reduced to the boundedness of the matrix series $\sum_{t=0}^{\infty} A^t$, and the stability in perturbation was reduced to the boundedness of the matrix sequence $M_t = \{A^t\}_{t=1}^{\infty}$.

We formulate and justify the following stability criterion for perturbation and value.

Criterion 2. The system in the form of signed weighted directed graph G with the adjacency matrix A is stable by perturbation IFF the spectral radius of the adjacency matrix is $\mathrm{p}(A) = \max_i |\lambda_i| \le 1$, where $\{\lambda_i\}_{i=1}^{M}$ are the eigenvalues A, and is the basis of eigenvectors.

We give a proof of this criterion. In accordance with (6.9), perturbation stability occurs because matrix sequence $M_t = \{A^t\}_{t=1}^{\infty}$ is limited. Let V_J be the Jordan basis A, then $A = V_J^{-1} A_J V_J$, where A_J is the Jordan form A. We write the matrix A in the Jordan basis (Gantmakher 1967)

$$A_J = \begin{bmatrix} J_1 & 0 & \cdots & 0 \\ 0 & J_2 & \cdots & 0 \\ \vdots & \vdots & \ddots & \vdots \\ 0 & 0 & \cdots & J_m \end{bmatrix}$$

where $J_1, \ldots, J_m, i = 1, 2, \ldots, m$ are the Jordan cells of dimension of the $m \times m$ form

$$J_i = \begin{bmatrix} \lambda_i & 1 & 0 & \cdots & 0 \\ 0 & \lambda_i & \ddots & \ddots & \vdots \\ \vdots & \ddots & \ddots & 1 & 0 \\ 0 & \cdots & 0 & \lambda_i & 1 \\ 0 & \cdots & 0 & 0 & \lambda_i \end{bmatrix}$$

corresponding to elementary divisors $(\lambda - \lambda_1)^{p_1}, \ (\lambda - \lambda_2)^{p_2}, \ \ldots, \ (\lambda - \lambda_m)^{p_m}$, $(p_1 + p_2 + \cdots + p_m = n)$. Then

$$A_J^t = \begin{bmatrix} J_1^t & 0 & \cdots & 0 \\ 0 & J_2^t & \cdots & 0 \\ \vdots & \vdots & \ddots & \vdots \\ 0 & 0 & \cdots & J_m^t \end{bmatrix} \tag{6.10}$$

It follows that boundedness $\left\{A^t\right\}_{t=1}^{\infty}$ is equivalent to boundedness $\left\{J_1^t\right\}_{t=1}^{\infty}$ for all Jordan cells of matrix A.

According to (6.10), for a Jordan cell, we have

$$J^t = \begin{bmatrix} \lambda^t & \frac{t\lambda^{t-1}}{1\,!} & \frac{t(t-1)\lambda_i^{t-2}}{2\,!} & \cdots & \frac{t!\lambda^{t-n+1}}{(t-n+1)!(n-1)!} \\ 0 & \lambda_i^t & \ddots & \ddots & \vdots \\ \vdots & \ddots & \ddots & \frac{t\lambda^{t-1}}{1\,!} & \frac{t(t-1)\lambda^{t-2}}{2\,!} \\ 0 & \ddots & 0 & \lambda^t & \frac{t\lambda^{t-1}}{1!} \\ 0 & \cdots & 0 & 0 & \lambda^t \end{bmatrix} =$$

$$= \begin{bmatrix} \lambda^t & C_t^1\lambda^{t-1} & C_t^2\lambda^{t-2} & \cdots & C_t^{n-1}\lambda^{t-n+1} \\ 0 & \lambda^t & \ddots & \ddots & \vdots \\ \vdots & \ddots & \ddots & C_t^1\lambda^{t-1} & C_t^2\lambda^{t-2} \\ 0 & \cdots & 0 & \lambda^t & C_t^1\lambda^{t-1} \\ 0 & \cdots & 0 & 0 & \lambda^t \end{bmatrix},$$

where sequence $\{C_i^t\lambda^{t-i}\}_{t=0}^{\infty}$ $\forall i = 0, 1, ..., n-1$ should be limited. Then if $|\lambda| < 1$, then $|C_i^t\lambda^{t-i}| \underset{t\to\infty}{\longrightarrow} 0$ $\forall i = 0, 1, ..., n-1$, since the polynomial grows more slowly than the power function decreases. Therefore, sequences are not limited $\forall i = 0, 1, ..., n-1$; $|\lambda| = 1$, then

$$\begin{cases} |C_i^t\lambda^{t-i}| \underset{t\to\infty}{\longrightarrow} \infty, \ \forall i = 1, ..., n-1, \\ |C_i^t\lambda^{t-i}| = |\lambda^t| = 1, \quad i = 0. \end{cases}$$

Therefore, sequences $\{C_i^t\lambda^{t-i}\}_{t=0}^{\infty}$ will be bounded $\forall i = 0, 1, ..., n-1$, IFF $n = 1$, and the Jordan cell has dimension 1. It follows that the sequence $\{A^t\}_{t=1}^{\infty}$ is limited $\forall i = 0, 1, ..., n-1$, IFF all eigenvalues of the matrix A modulo less than 1 or do not exceed 1, and the Jordan form of the matrix is diagonal. Thus, the graph is stable in perturbation since $\rho(A) = \max_i|\lambda_i| \leq 1$, as required to proof.

Now we formulate a criterion for stability with respect to the initial value:

Criterion 2. The system in the form of signed weighted directed graph G with the adjacency matrix A is stable by value IFF the spectral radius of the adjacency matrix is $\rho(A) = \max_i|\lambda_i| < 1$, where $\{\lambda_i\}_{i=1}^{M}$ are characteristic numbers A, or $\rho(A) = 1$, but the Jordan form of the matrix is diagonal and there is no eigenvalue equal to 1.

We give a proof of this criterion. We write the matrix equality:

$$(I + A + A^2 + \cdots + A^t)(I - A) = I - A^{t+1} \tag{6.11}$$

We assume that the system is stable in value, i.e., the matrix series $\sum_{t=0}^{\infty} A^t$ is bounded (in norm) by the constant C. Then from the expression (6.10), we get

$$\begin{aligned} \|A^{t+1}\| &= \|A^{t+1} - I + I\| \leq \|A^{t+1} - I\| + \|I\| \\ &= \|(I + A + A^2 + \cdots + A^t)(I - A)\| + 1 \leq \\ &\leq \|(I + A + A^2 + \cdots + A^t)\| * \|(I - A)\| + 1 \leq C\|(I - A)\| + 1. \end{aligned}$$

Therefore, the sequence $\{A^t\}_{t=1}^{\infty}$ is bounded, whence from criterion 1 we obtain the first part of the necessary statement, i.e., $\rho(A) = \max_i|\lambda_i| < 1$.

It is also necessary to prove that there is no eigenvalue equal to 1. Assume the converse, that is $\exists x \neq 0 : \ Ax = x$. Then

$$\|(I + A + A^2 + \cdots + A^t)x\| = \|(t+1)x\| \underset{t\to\infty}{\longrightarrow} \infty,$$

$$\|(I + A + A^2 + \cdots + A^t)\| = \sup_{x \neq 0} \frac{\|(I + A + A^2 + \cdots + A^t)x\|}{\|x\|} \geq t$$
$$+ 1\|(t+1)x\| \underset{t\to\infty}{\longrightarrow} \infty.$$

The amount received is not limited. Therefore, an eigenvalue equal to 1 does not exist.

We assume that the spectral radius of the matrix A is 1, but the matrix has a basis from eigenvectors and 1 is not a characteristic number of the adjacency matrix. Then, by criterion 1, we obtain that the sequence $\{A^t\}_{t=1}^{\infty}$ is bounded by the constant C. In addition, since 1 does not belong to the spectrum of the matrix A, then 1 belongs to the resolvent set, that is $\exists(I - A)^{-1}$. . Then, after multiplying both sides of (6.6) on $(I - A)^{-1}$ from the right, we get

$$I + A + A^2 + \cdots + A^t = (I - A^{t+1})(I - A)^{-1}$$

From here

$$\begin{aligned}\left\|(I + A + A^2 + \cdots + A^t)\right\| &= \left\|(I - A^{t+1})(I - A)^{-1}\right\| \\ &\leq \left\|(I - A^{t+1})\right\| * \left\|(I - A)^{-1}\right\| \\ &\leq (\|I\| + \left\|A^{t+1}\right\|) * \left\|(I - A)^{-1}\right\| \leq (1 + C) * \left\|(I - A)^{-1}\right\|.\end{aligned}$$

Therefore, the matrix series $\sum_{t=0}^{\infty} A^t$ is limited. Thus, the graph is stable by the value of t.t.c. $\rho(A) = \max_i|\lambda_i| < 1$, or, $\rho(A) = 1$ as required to prove. Note that a graph is numerically stable if the spectral radius of the adjacency matrix is less than 1.

The study of structural stability is carried out by analyzing all cycles of graph G. Among the cycles, cycles of positive and negative feedback are distinguished. A positive feedback cycle is a cycle in which there are no or an even number of negative arcs; this is a cycle of processes accelerator in the system. A negative feedback cycle is a cycle in which there is an odd number of negative arcs; this is a process stabilizer cycle. A condition for the structural stability of the system is the presence of an odd number of negative cycles in it (Roberts 1978).

Investigations of the connectivity and complexity of the system are necessary for solving problems about the possibility of managing the system, selecting management methods, assessing the conditions necessary for the implementation of management.

The analysis of the connectivity of the system under cognitive modeling can be carried out both on the basis of graph theory and using the language of algebraic topology, which makes it possible to analyze the structure of a complex multidimensional geometric formation, the simplicial complex. This is possible with minimal a priori information regarding the objects and phenomena under study. The mathematical foundations for the polyhedral (simplicial) analysis were laid by C. Droucer, and further analysis was obtained in the works of the British physicist R. H. Atkin (Atkin and Casti 1977; Atkin 1997).

To perform the simplicial analysis, it is necessary to isolate the simplexes generated by each vertex of the cognitive model to determine the chains of bonds (q-connections) between them and to construct simplicial complexes.

The loosely coupled system allows for offline block management. This can have both positive and negative sides. The negative effect arises from the fact that autonomous management within a real unified social and economic system can prove to be not only costly (fragmentation of funds) but also harmful because of the non-observance of the principle of systemic nature. A tightly coupled system with a rigid structure along with the advantages of an "easier" centralized control is less flexible and effective in a rapidly changing environment, it is more difficult to adapt to dynamic changes. Under such conditions, systems with a tunable structure may be more effective. But in any case, it is necessary to study the structural properties of the existing complex system.

The third step. Scenario analysis, pulse modeling. At the third stage of cognitive modeling, to determine the possible development of processes in a complex system and develop development scenarios, we used the impulse process model (modeling the propagation of disturbances in cognitive models), which is a transition from a model (6.6) to a model of steps $t_n \to n$ (Roberts 1978):

$$x_i(n+1) = x_i(n) + \sum_{i=1}^{k-1} f(x_i, x_j, e_{ij}) P_j(n) + Q_i(n) \quad (6.12)$$

where $x(n)$, $x(n+1)$ are the values of the indicator at the vertex V_i at the modeling steps at time $t =$ n and the next $t = n + 1$; $P_j\ (n)$ is the pulse existed at the vertex V_j at the moment $t = n$; $Q_{Vi}\ (n) = \{q_1, q_2, ..., q_k\}$ is the vector of external pulses (disturbing or controlling actions) introduced to the vertexes V_i at time moment n. Scenario analysis is carried out by the means of an impulse simulation. To generate possible scenarios of the system development at the vertices of the cognitive map, hypothetical perturbing/control actions (impulses) $Q = \{q_{ij}\}$ are introduced. A set of implementations of impulse processes is a "scenario of development," which indicates possible trends in the development of situations in the system. The situation in impulse simulation is characterized by a set of all perturbing effects of Q and X values at each step of the simulation. The scenario answers the question: "What will happen if …?". The development scenario is one of the hypothetical variants of the future processes in system—the foresight of the future.

The completion of cognitive modeling should be the choice of the desired scenario for the development of the system, the development, and justification of management decisions are aimed at implementing the desired scenario, preventing the consequences of unwanted scenarios.

If the decision-maker is not satisfied with the results of cognitive modeling, model adjustment is necessary. It is possible to change the number of vertices and the relationships between them, rearranging the structure of the initial cognitive model.

6.2 Applying Cognitive Modeling to Urban Underground Construction

6.2.1 Description of the Urban Underground Construction Problem

Models and methods of foresight and cognitive modeling of complex systems allow us to develop a cognitive model of the system, use it to analyze the structural properties of the system, its stability, and, most importantly, analyze the possible ways of developing the system taking into account changes of its internal and external environment parameters and under various control actions. Let us consider in which direction the study of the problems of construction geotechnology can be used for foresight and cognitive modeling of complex systems.

Currently, the structure of the science for building geotechnology is presented in four main sections: underground urbanism, mechanics of underground structures, geonics, and rock management during construction. For research using the proposed cognitive methodology, the content of all sections is of interest. So, the content of one of the directions of the first section is the substantiation of strategies and methods for developing urban underground space, which is also one of the tasks of cognitive modeling complex systems.

At the design stage of underground structures, it is necessary to consider and justify the practical necessity, socio-economic expediency, and technical feasibility of constructing underground structures in mining and geological conditions and under the influence of construction technology, the functional purpose of construction objects. The tasks of substantiating the necessity, feasibility, expediency, and effectiveness of actions in complex systems are also included in the field of cognitive modeling tasks. The knowledge and data from the other scientific sections of construction geotechnology can also be useful in cognitive modeling of the structure and behavior of a system that imitates a real system from the perspective of the stated goal of research and functioning.

So, "assessment of the stability of mine workings; the study of the processes of engineering structures interaction with rock masses and the establishment of qualitative and quantitative characteristics of their stress-strain state" and others (second section) can provide data and gain new knowledge about them after cognitive modeling when analyzing of the cognitive model properties; the third section, geonics, including research on the "interconnections of the elements of mining construction technology, the establishment of qualitative and quantitative parameters that determine the choice of methods, engineering and construction technology; effective methods of organizing labor and managing construction works …" turn out to be necessary when developing a cognitive model, establishing relationships between its objects (or "concepts", "factors", and "entities"), which are also concepts (concepts) associated with urban underground planning.

A significant advantage of cognitive models is that their composition at different stages and levels of study and description can include both quantitative (for example, "hydrostatic pressure") and qualitative (for example, "socio-economic feasibility") characteristics.

In our opinion, imitation of cognitive modeling, especially at the stage of design work on the development of the underground space, is extremely necessary. A serious reason for this may be the fact that it is necessary to anticipate and eliminate or reduce risks in advance (manage risks), which are inevitably inherent in urban underground planning. As shown in Levchenko (2007): "Urban underground construction is characterized by dynamism and a high degree of uncertainty, so the risk factor is an integral attribute of the development of underground space". So, risks under certain conditions are manifested and may have negative consequences for the entire underground infrastructure in the system "man-underground construction-environment" (Saługa 2009; Kulikova et al. 2005).

Knowledge of risks is necessary for everyone: designers, builders, and operators. In (Kulikova et al. 2005), a risk classification was proposed, consisting of eight groups: construction, environmental, managerial and executive, commercial, economic, contractual, social, and operational. The classification provides the basis for the further development of environmentally friendly technologies and construction methods. Construction risk is dominant in terms of the impact on the entire life cycle of an underground structure and the higher it is, the lower the requirements for personnel qualifications, construction quality and terms, reliability of mining equipment, etc.

Wrong construction decisions are the basis for the occurrence of environmental, economic, operational, and other risks. Therefore, the basic principle laid down in research to improve the design and construction methods of underground facilities is the principle of minimizing damage from the consequences of negative manifestations of risks, taking into account the interaction and mutual influence of all natural, technical, technological, and other factors. The possibility of disasters caused not only by unpredictable natural situations but also by design errors and imperfection of existing technologies requires the formulation and solution of the problem of the viability of the object in extreme and emergency situations.

Therefore, the goal of the cognitive modeling presented in this paper was to study some problems of the object's viability in extreme and emergency situations. The system approach to solving of this problem required the definition and description of the main elements (parameters, factors, concepts), causes, and effects characterizing the natural and technical geosystem ("underground structure-environment"). As a result of cognitive modeling, scenarios of the possible development of a complex system that may arise under the influence of changes in the internal and external environment of the underground structure should be obtained. This is especially important for knowing and preventing negative consequences, minimizing damage under the influence of the most unfavorable combination of negative factors: external and internal static and dynamic loads, all kinds of man-made influences inside the underground structure, harmful natural manifestations from the rock mass, etc.

Ultimately, the task boils down to the problem of risk management as to the task of choosing options for a strategy and tactics of action (in relation to underground construction, these are constructive solutions, technology, and organization of work) that allow partially or completely avoiding risks, limiting their number and, ultimately, minimize damage from their negative manifestations.

6.2.2 Modeling of Urban Underground Construction

Consider the process of modeling urban underground construction without reference to a specific area, in a generalized form (Pankratova and Pankratov 2022).

First step. Cognitive Model Development. Table 6.1 presents data on the vertices (concepts) of the hierarchical cognitive model without reference to a specific territory. These data are obtained using the SWOT analysis method. We used generalizing concepts (indicators, factors), independent of the specifics, which can be disclosed and taken into account in the future when developing the lower levels of the hierarchical model, are using. Figure 6.2 shows a hierarchical cognitive map I_G: "Natural-technical geosystem". In Table 6.1 and Fig. 6.2, the vertices of the upper (first level) are denoted as $I - V_i$, $i = 5, 11, 13, 15, 16$. In Table 6.1, in the column "vertex assignment", vertices that play a different roles in the cognitive system are highlighted.

Figure 6.3 represents the fragment of the cognitive map I_G relationship matrix A_G (6.2).

The cognitive model is a simulation model that makes it possible not to conduct an experiment on a "living" system, but to simulate its behavior and possible future development under the influence of various factors, generating new knowledge about the system. This allows you to justify management decisions in a given situation.

Second step of cognitive modeling. Before using the cognitive model to determine its possible behavior, the second stage of modeling analyzes the various properties of the model is realized. In this case, the stability properties of the model must be analyzed.

Figures 6.4 and 6.5 show an example of determining the ycles of the cognitive model I_G.

Figure 6.4 shows one of the positive feedback cycles, a sign of which is an even number of negative arcs in it. Figure 6.5 presents one of the negative cycles.

Impulse sustainability. The cognitive model I_G was not resistant to perturbations according to the accepted criterion (Roberts 1978; Kulba et al. 2002): the maximum modulo M root of the characteristic equation of the matrix of relations of the graph I_G is $|M| = 1.82 > 1$ (must be less than 1).

Structural stability. An analysis of the ratio of the number of stabilizing cycles (35 negative feedbacks) and process accelerator cycles (33 positive feedbacks) indicates the structural stability of such a system (Roberts 1978).

Table 6.1 The vertices of the hierarchical cognitive map "Natural-technical geosystem"

Code	Vertex explanation	Vertex assignment
$I-V_{11}$	The viability of the urban underground development	Indicative
$I-V_{13}$	Disasters, extreme and emergency situations	Perturbing
$I-V_{15}$	Environmental risks	Perturbing
$I-V_{16}$	Economic risks	Perturbing
$I-V_5$	Genetic type and lithological composition of soils	Basic
V_1	Mountain and hydrostatic pressure, seismic impact	Basic
V_2	Surface Load Static Load Index	Basic
V_3	The indicator of the static load of the surrounding soil massif	Basic
V_4	Existing underground facilities	Disturbing
V_6	Estimated soil resistance	Basic
V_7	Aquifers and High Water	Disturbing
V_8	Relief Type and Morphometry	Basic
V_9	Engineering and geological processes	Disturbing
V_{10}	Mining construction technologies	Regulating
V_{12}	The level of comfort of work and rest during the construction and operation of underground structures	Indicative
V_{14}	Construction, operational, and management risks	Disturbing
V_{17}	Staff qualifications	Regulating
V_{18}	Industrial Safety	Basic
V_{19}	Quality and construction time	Regulating

The given example of the analysis of the cycles of the cognitive model showed the variety of cycles of cause-and-effect relationships that exist in complex systems. There are 68 of them in the analyzed system. Without an appropriate theoretical analysis, there is a great risk of the human factor in making managerial decisions, because its consequences may not be obvious due to the complexity of interactions in the system.

Analysis of system connectivity, simplicial analysis. Immersed in the study of the structure of the cognitive model, it is desirable to conduct a simplicial analysis of the properties of its connectivity. Such an analysis is carried out in order to study and understand the topological properties of the model and, accordingly, other connectivity faces of the complex system under study that are not detected in the algebraic analysis. According to R.H. Atkin and J. Casti, connectedness is the essence of the concept of a large system (Atkin and Casti 1977; Atkin 1997; Casti 1979).

The connectivity properties of blocks (simplexes) characterize the "deep" connections of the cognitive model, the connections of its simplexes, and not just the vertices, as in the cognitive map. A simplex is formed by each vertex, which is the reason that some other vertices interact with each other. Figures 6.6, 6.7, and 6.8 show the

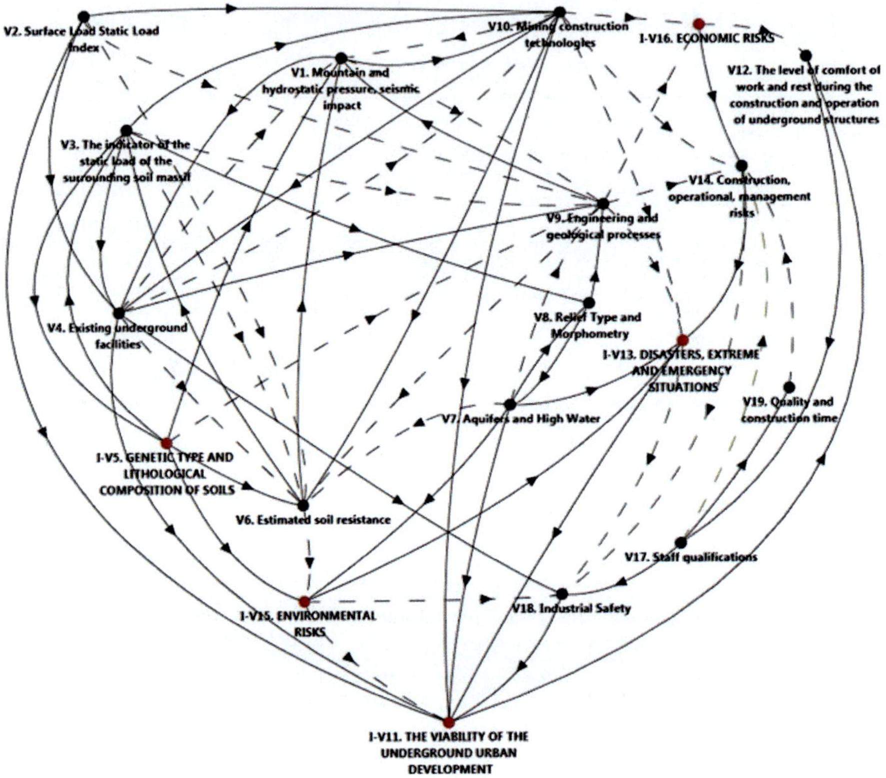

Fig. 6.2 Hierarchical cognitive map I_G "Natural-technical geosystem"

Connectivity matrix

- Infl./ Dep.	V1	V2	V3	V4	I-V5	V6	V7	V8	V9	V10	I-V11	V12	I-V13	V1
V1	X			1.0					-1.0	1.0				
V2		X		1.0		-1.0			-1.0	1.0	1.0			
V3			X	1.0	1.0	-1.0		1.0	-1.0	1.0				
V4	-1.0			X		-1.0			1.0	-1.0	1.0			
I-V5	1.0		1.0		X	1.0			-1.0					
V6	1.0		1.0			X								
V7						-1.0	X	1.0	-1.0		1.0		1.0	
V8							1.0	X	1.0					
V9	1.0					-1.0			X				-1.0	-1.0

Fig. 6.3 The fragment of the cognitive map I_G relationship matrix A_G

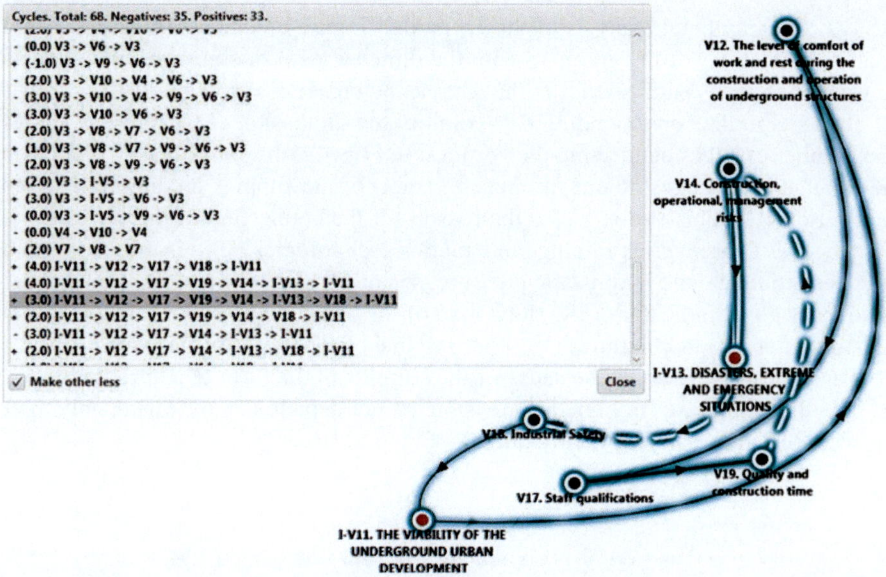

Fig. 6.4 Cognitive map cycles, one of the positive cycles is highlighted

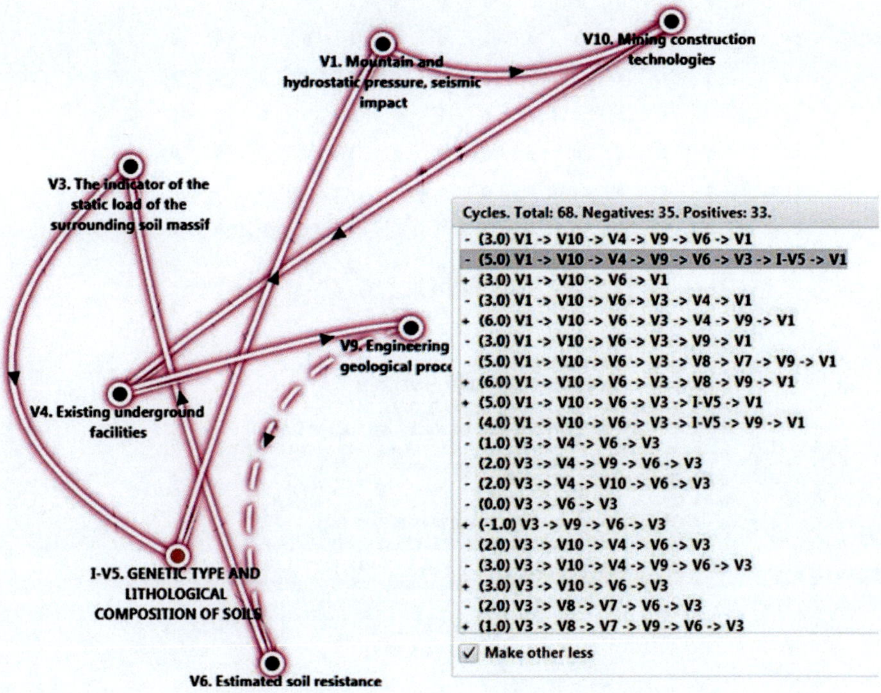

Fig. 6.5 Cognitive map cycles, one of the negative cycles is highlighted

results of a simplicial analysis of the I_G model. Figure 6.6 shows the transformed matrix of relations of the graph I_G with the dimensions ρ of simplexes $\sigma_\rho^{V_i}$ of rows (x) and columns (y) indicated it in the decreasing order; $\rho = k - 1$, k is the number of elements in the corresponding row/column, the dimension of the simplex shows the number of edges connecting the vertices. In Fig. 6.7, the simplices $\sigma_{\rho=3}^{(V_{10})}$ ($\rho = 3$ means that three edges go out from each vertex) of the simplex are highlighted for one vertex V_{10}. This vertex V_{10} is the reason for the connection of the vertices V_1, V_2, V_3, V_4. Those, vertex mining construction technologies (V_{10}) is the reason that vertices mountain and hydrostatic pressure, seismic impact (V_1), surface load static load index (V_2), indicator of the static load of the surrounding soil massif (V_3), and existing underground facilities (V_4) forming one block are interconnected.

Thus, these vertices are the cause of the simplex in the form of a tetrahedron.

Note that simplexes of higher dimension are not depicted on the plane; only their "projection" can be conditionally drawn—Fig. 6.8.

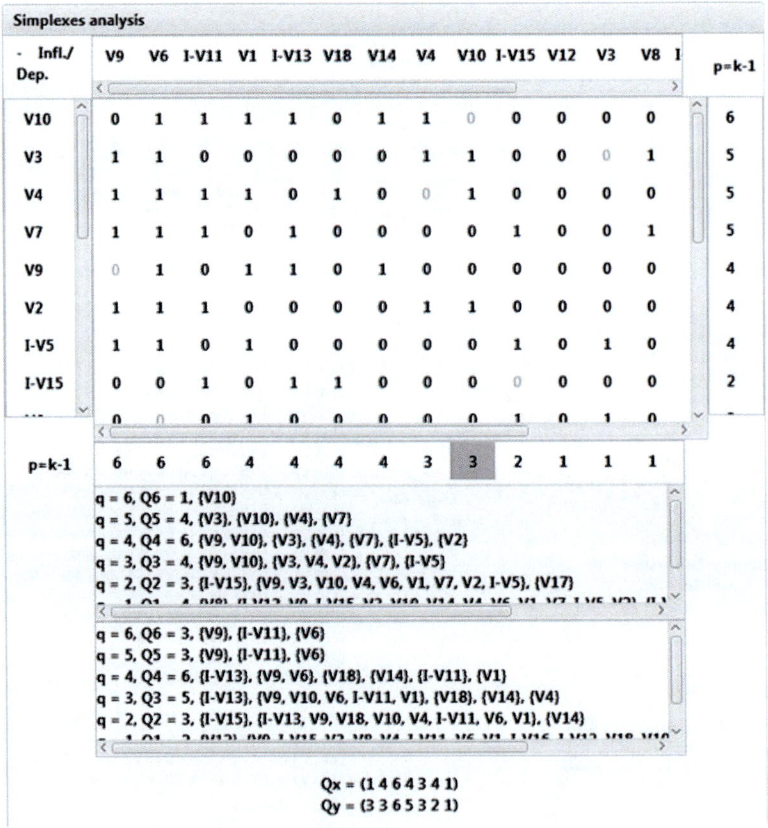

Simplexes analysis

- Infl./ Dep.	V9	V6	I-V11	V1	I-V13	V18	V14	V4	V10	I-V15	V12	V3	V8	p=k-1
V10	0	1	1	1	1	0	1	1	0	0	0	0	0	6
V3	1	1	0	0	0	0	0	1	1	0	0	0	1	5
V4	1	1	1	1	0	1	0	0	1	0	0	0	0	5
V7	1	1	1	0	1	0	0	0	0	1	0	0	1	5
V9	0	1	0	1	1	0	1	0	0	0	0	0	0	4
V2	1	1	1	0	0	0	0	1	1	0	0	0	0	4
I-V5	1	1	0	1	0	0	0	0	0	1	0	1	0	4
I-V15	0	0	1	0	1	1	0	0	0	0	0	0	0	2
p=k-1	6	6	6	4	4	4	4	3	3	2	1	1	1	

q = 6, Q6 = 1, {V10}
q = 5, Q5 = 4, {V3}, {V10}, {V4}, {V7}
q = 4, Q4 = 6, {V9, V10}, {V3}, {V4}, {V7}, {I-V5}, {V2}
q = 3, Q3 = 4, {V9, V10}, {V3, V4, V2}, {V7}, {I-V5}
q = 2, Q2 = 3, {I-V15}, {V9, V3, V10, V4, V6, V1, V7, V2, I-V5}, {V17}

q = 6, Q6 = 3, {V9}, {I-V11}, {V6}
q = 5, Q5 = 3, {V9}, {I-V11}, {V6}
q = 4, Q4 = 6, {I-V13}, {V9, V6}, {V18}, {V14}, {I-V11}, {V1}
q = 3, Q3 = 5, {I-V13}, {V9, V10, V6, I-V11, V1}, {V18}, {V14}, {V4}
q = 2, Q2 = 3, {I-V15}, {I-V13, V9, V18, V10, V4, I-V11, V6, V1}, {V14}

Qx = (1 4 6 4 3 4 1)
Qy = (3 3 6 5 3 2 1)

Fig. 6.6 Results of simplicial analysis (calculation)

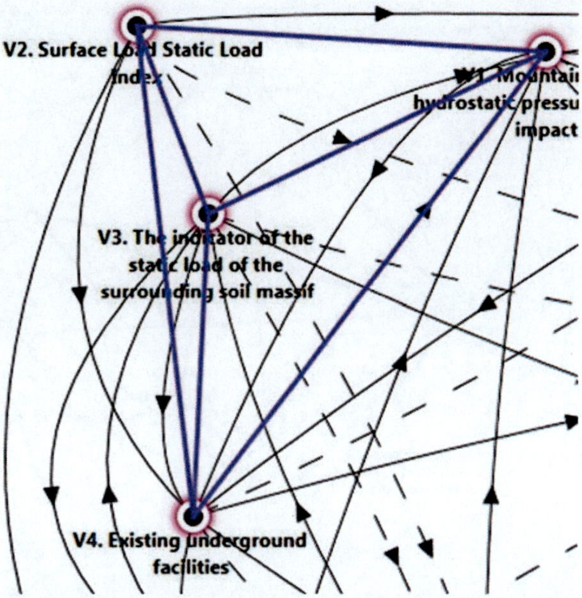

Fig. 6.7 Image of one of the simplices of dimension $\rho = 3$

Simplexes form q-connected chains $q = 1$ (connection through a vertex), $q = 2$ (connection through an edge), $q = 3$ (connection through a plane), etc., thus uniting into simplicial complexes K_x (along the lines—"inputs") and K_y (columns—"outputs"). Simplexes are q-connected or not in simplexes (they lack or have connections of simplexes along vertices, edges, planes, m-dimensional volumes). Simplicial complexes are characterized by the structural vectors Q_x and Q_y formed by vertex groups common to different simplexes.

The third step of modeling. Scenario analysis is designed to anticipate possible trends in the development of situations on the model. To generate scenarios of the development of the system, impacts are introduced into the vertices of the cognitive map in the form of a set of impulses. The impulse process formula has the form (6.12).

Let us introduce perturbations Q of different sizes (normalized) to any of the vertices, as well as to their combination. In connection with a large number of theoretically possible variants of introduced disturbances, it is expedient to develop a plan for a computational experiment before excluding pulse simulation, eliminating at least almost impossible variants. Introducing disturbances to the vertices, the decision-maker is looking for the answer to the question: "What will happen if …?".

The software system allows, in the process of impulse modeling and analysis of the obtained results, to introduce control or disturbing influences at any modeling step. It allows to change (correct) scenarios in model dynamics, to determine the effects that bring the processes closer to the desired.

Let us present the results of pulse modeling in the third scenario.

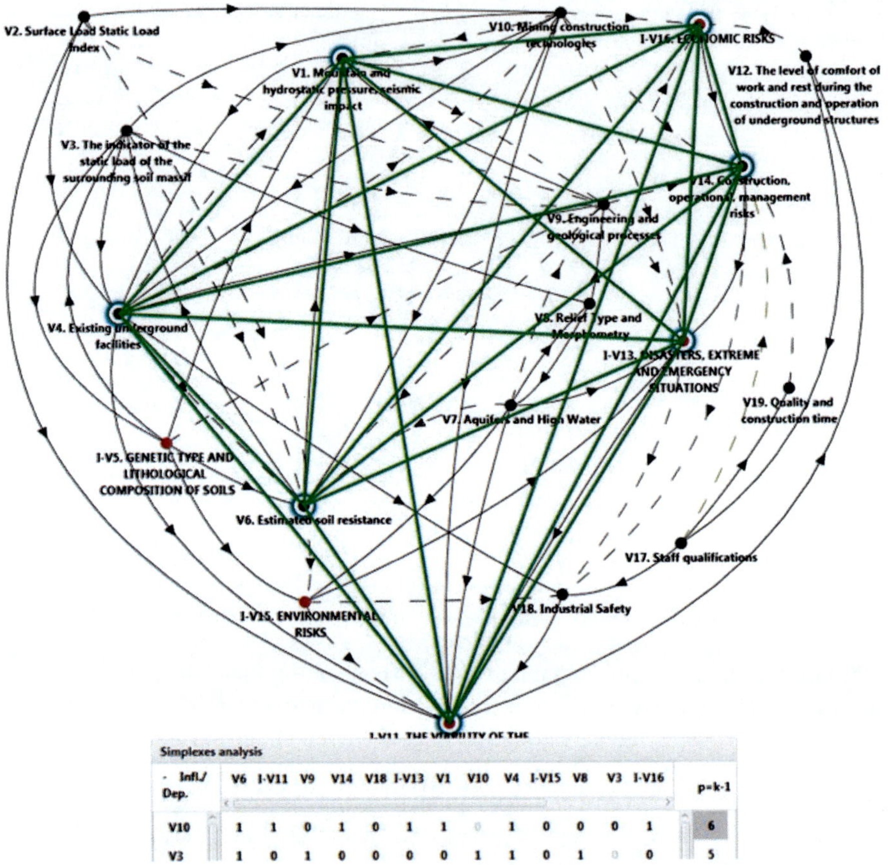

Infl./ Dep.	V6	I-V11	V9	V14	V18	I-V13	V1	V10	V4	I-V15	V8	V3	I-V16	p=k-1
V10	1	1	0	1	0	1	1	0	1	0	0	0	1	6
V3	1	0	1	0	0	0	0	1	1	0	1	0	0	5

Fig. 6.8 Image of the projection of one of the simplexes of dimension $\rho = 6$

Scenario No. 1. Assume good technology is used in underground construction. To the vertex V_{10}, the control action is introduced $q_{10} = +1$, the perturbation vector $Q = \{q_1 = 0, \ldots q_{10} = +1, \ldots q_{19} = 0\}$.

Table 6.2 presents the results of a computational experiment, 10 simulation steps for Scenario No. 1.

Figures 6.9 and 6.10 show graphs of pulsed processes constructed according to the data in Table 6.2. For the convenience of visual analysis of the image, the graphs of pulsed processes in the vertices V_{10}, $I - V_{15}$, $I - V_{16}$, $I - V_{11}$, and $I - V_5$ are represented by two figures: Fig. 6.9—from the first to the sixth steps of modeling and Fig. 6.10 from the sixth to the tenth steps of modeling. The image of pulsed processes at a larger number of simulation steps is not necessary, because system behavior trends under these conditions are already evident.

Modeling scenario No. 1, it is advisable to analyze whether changes in mining construction technologies (V_{10}) can and in what way affect other vertices of the

Table 6.2 The results of a computational experiment. Scenario No. 1

Step Vertex	0.0	1.0	2.0	3.0	4.0	5.0	6.0	7.0	8.0	9.0	10.0
VI. Mountain and hydrostatic pressure, seismic impact	0.0	0.0	1.0	1.0	3.0	−3.0	−10.0	23.0	8.0	−98.0	94.0
V2. Surface Load Static Load Index	0.0	0.0	0.0	0.0	0.0	0.0	0.0	0.0	0.0	0.0	0.0
VB. The indicator of the static load of the surrounding soil massif	0.0	0.0	0.0	1.0	0.0	−3.0	3.0	4.0	−20.0	9.0	64.0
V4. Existing underground facilities	0.0	0.0	1.0	0.0	−1.0	4.0	*1.0	−16.0	22.0	32.0	−122,0
I−VS. GENEIIC TYPE AND L1THOLGG1CAL COMPOSITION OF SOILS	0.0	0.0	0.0	0.0	1.0	0.0	−3.0	3.0	4.0	20.0	9.0
V6. Estimated soil resistance	0.0	0.0	1.0	0.0	−4.0	3.0	7.0	*23.0	5.0	84.0	−125.0
V7. Aquifers and High Water	0.0	0.0	0.0	0.0	0.0	1.0	0.0	−2.0	3.0	2.0	−17.0
V8. Relief Type and Morphometry	0.0	0.0	0.0	0.0	1.0	0.0	−2.0	3.0	2.0	−17.0	11.0
V9. Engineering and geological processes	0.0	0.0	0.0	2.0	0.0	4.0	9.0	7.0	41.0	29.0	122,0
V10. Mining construction technologies	0.0	1.0	1.0	−1.0	1.0	5.0	−9.0	−5.0	44.0	−33.0	−120.0
1-VLL. THE VIABILITY OF THE UNDERGROUND URBAN DEVELOPMENT	0.0	0.0	1.0	1,0	1.0	1.0	9.0	1.0	−13.0	35,0	50.0
VI2. The level of comfoit of work and rest during the construction and operation of underground structures	0,0	0.0	0.0	2.0	2.0	2.0	2.0	10–0	1.0	−11.0	38.0

(continued)

Table 6.2 (continued)

Step Vertex	0.0	1.0	2.0	3.0	4.0	5.0	6.0	7.0	8.0	9.0	10.0
I-V13. DISASTERS, EXTREME AND EMERGENCY SITUATIONS	0.0	0.0	−1.0	‘2.0	‘4.0	‘3.0	1.0	−8.0	*19.0	18.0	•9.0
V14. Construction, operational, management risks	0.0	0.0	−1.0	−2.0	−2.0	−4.0	−6.0	−5.0	−6,0	−17.0	−10.0
I-V15. ENVIRONMENTAL RISKS	0.0	0.0	0.0	1.0	0.0	5.0	−2.0	10.0	24.0	2.0	−102.0
I-V16. ECONOMIC RISKS	0.0	0.0	1.0	1.0	−10	−10	−10	0.0	−2.0	−3.0	4.0
VI7. Staff qualifications	0.0	0.0	0.0	0.0	2.0	2.0	2.0	2.0	10.0	1.0	−11.0
VI8. Industrial Safety	0.0	0.0	0.0	3.0	5.0	7.0	8.0	8.0	9.0	33.0	30.0
VIS. Quality and construction time	0.0	0.0	0.0	0.0	0.0	2.0	2.0	2.0	2.0	10.0	1.0

cognitive model. As can be seen from the graphs in Figs. 6.9 and 6.10, positive changes in V_{10} can contribute to positive trends in the development of vertices at the top hierarchical level: up to the fifth and the sixth steps of the modeling, the declining trends of disasters, extreme and emergency situations ($I - V_{13}$), environmental risks ($I - V_{15}$), economic risks ($I - V_{16}$), genetic type and lithological composition of soils ($I - V_5$), and viability of the urban underground development ($I - V_{11}$) is growing. Further, the oscillatory mode manifests itself more and more in changing situations, as can be seen from the graphs in Fig. 6.11.

All this may indicate that a single positive change in one of the vertices of the system model may not be enough to exclude the negative impact of risks and other negative influences.

Scenario No. 2. Suppose that the possibility of the simultaneous occurrence of all risks is increasing in the system. Disturbing effects are appearing $q_{14} = +1$, $q_{15} = +1$, $q_{16} = +1$, there is a perturbation vector $Q = \{q_1 = 0, \ldots q_{14} = +1, q_{15} = +1, q_{16} = +1, \ldots q_{19} = 0\}$.

Pulse simulation results are presented in Table 6.3, Fig. 6.12 for vertices V_{14}, $I - V_{15}, I - V_{16}, I - V_{11}, I - V_{13}, V_{17}, V_{18}, V_{19}$ and Fig. 6.13 for vertices $V_{12}, V_{17}, V_{18}, V_{19}, V_{10}$.

The simulation results of the second scenario show an extremely unfavorable option for the development of situations in the system. With increasing risks, as can be seen from Table 6.3 and Figs. 6.12 and 6.13, all indicators of the system fall at both the first and second levels of the hierarchy. This observation forces one to make a decision on the search for the necessary counteraction to the situations that have arisen.

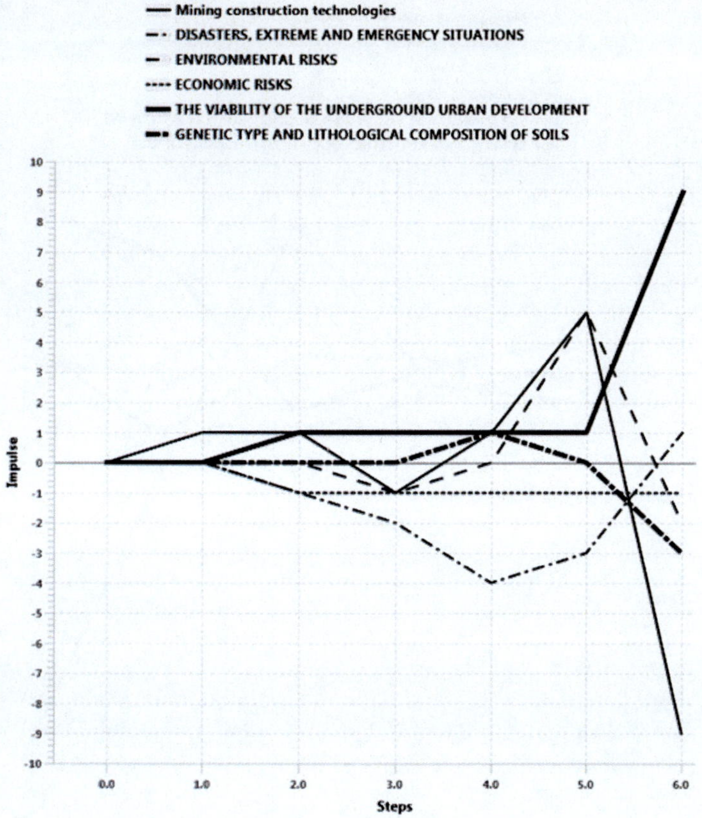

Fig. 6.9 Graphs of pulsed processes, from the first to the sixth steps of modeling. Scenario No. 1

Consider the third scenario. Suppose improving engineering and geological processes (V_9), mining construction technologies (V_{10}), staff qualifications (V_{17}), and quality and construction time (V_{19}), but there are disasters, extreme and emergency situations ($I - V_{13}$).

Scenario No. 3. Control actions $q_9 = +1, q_{10} = +1, q_{17} = +1, q_{19} = +1, q_{13} = +1$, the perturbation vector $Q = \{q_1 = 0, \ldots q_9 = +1, q_{10} = +1, \ldots q_{13} = +1, \ldots q_{17} = +1, \ldots q_{19} = +1\}$.

The results of pulse modeling are presented in Table 6.4 and Fig. 6.14 for vertices $I - V_{13}, V_9, V_{10}, V_{19}, I - V_{15}, I - V_{16}, V_{18}, V_{17}, I - V_{11}$ and Fig. 6.15 for vertices $V_{17}, V_{18}, V_{19}, I - V_{11}, V_{12}, V_{14}, I - V_{13}$.

An analysis of the results of impulse modeling according to scenario No. 3 shows that the introduction of control actions to the vertices of engineering and geological processes (V_9), mining construction technologies (V_{10}), staff qualifications (V_{17}), and quality and construction time (V_{19}), but there are disasters, extreme and emergency situations ($I - V_{13}$) can counteract the negative impact of possible disasters

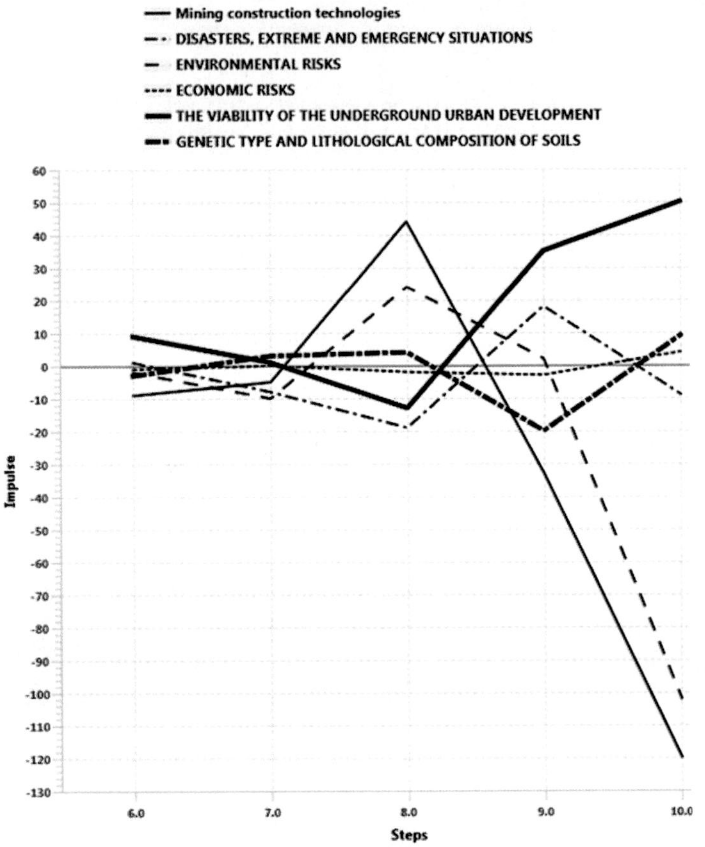

Fig. 6.10 Graphs of pulsed processes, from the sixth to the tenth step of modeling. Scenario No. 1

and extreme situations, reducing the impact of economic, environmental, and technological risks. Thus, scenario No. 3 can be considered favorable: industrial safety is increasing.

We present the simulation results in one more scenario No. 4. Assume that construction, operational, and management risks can be reduced. In this case, the impulse actions initiate six vertices of the model and the synergistic effect of their joint action is investigated. The modeling of this scenario of the situation's development on the model is carried out in order to determine whether it is necessary or not to strengthen the impact on the system to achieve good indicators.

Scenario No. 4. Control actions $q_9 = +1$, $q_{10} = +1$, $q_{17} = +1$, $q_{19} = +1$, $q_{13} = +1$, $q_{14} = -1$, the perturbation vector $Q = \{q_1 = 0, \ldots q_9 = +1, q_{10} = +1, \ldots q_{13} = +1, q_{14} = -1, \ldots q_{17} = +1, \ldots q_{19} = +1\}$.

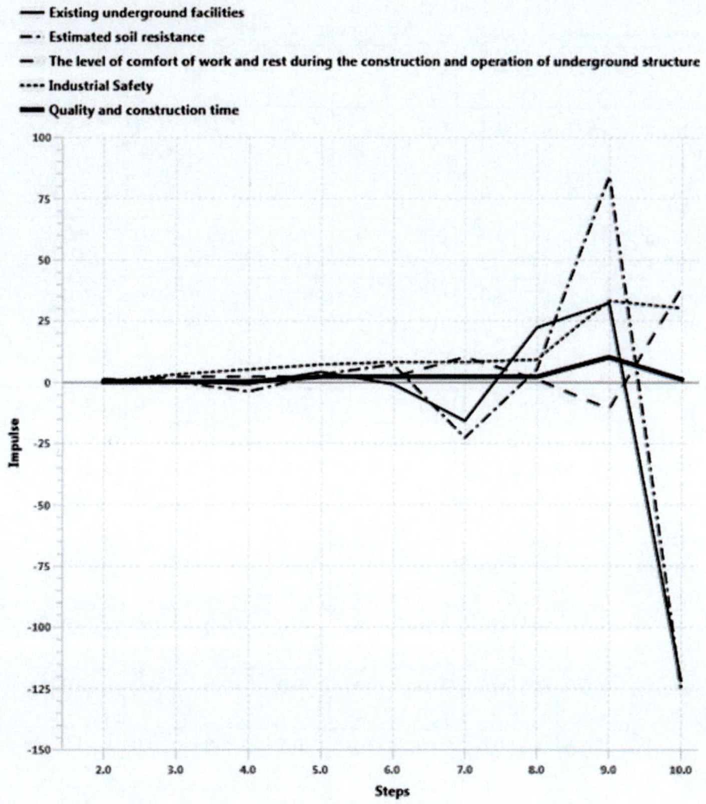

Fig. 6.11 Graphs of pulsed processes, vertices V_4, V_6, V_{12}, V_{18}, V_{19}. Scenario No. 1

The results of pulse modeling are presented in Table 6.5 and Fig. 6.16 for vertices $I - V_{13}$, V_9, V_{10}, V_{17}, V_{19} and Fig. 6.17 for vertices V_{12}, V_{14}, V_{15}, V_{16}, V_{18}, $I - V_{11}$.

Analysis of the simulation results of Scenario No. 4, which differs from scenario No. 3 by the addition of an impulse $q14 = -1$, simulating the possibility of reducing construction, operational, and management risks showed the following. The combined positive impact of six factors on the system leads to the possibility of the appearance of desirable trends in situations throughout the system. So, there are tendencies of improvement (growth) of the urban underground development viability, the level of comfort, work and rest during the construction and operation of underground structures, industrial safety while reducing all types of risk and reducing Disasters, extreme and emergency situations.

Let us compare the simulation results of scenarios No. 1, No. 2, No. 3, and No. 4, using the capabilities of the software system. We select the results of pulse modeling at the tenth step of modeling and present them in the form of histograms in Figs. 6.18 – 6.21.

Table 6.3 The results of a computational experiment. Scenario No. 2

Step Vertex	0.0	1.0	2.0	3.0	4.0	5.0	6.0	7.0	8.0	9.0	10.0
VI. Mountain and hydrostatic pressure, seismic impact	0.0	0.0	0.0	0.0	0.0	0.0	0.0	0.0	0.0	0.0	0.0
V2. Surface Load Static Load Index	0.0	0.0	0.0	0.0	0.0	0.0	0.0	0.0	0.0	0.0	0.0
V3. The Indicator of the static load of the surrounding soil massif	0.0	0.0	0.0	0.0	0.0	0.0	0.0	0.0	0.0	0.0	0.0
V4. Existing underground facilities	0.0	0.0	0.0	0.0	0.0	0.0	0.0	0.0	0.0	0.0	0.0
I-V5. GENETIC TYPE AND LITHOLOGICAL COMPOSITION OF SOILS	0.0	0.0	0.0	0.0	0.0	0.0	0.0	0.0	0.0	0.0	0.0
V6. Estimated soil resistance	0.0	0.0	0.0	0.0	0.0	0.0	0.0	0.0	0.0	0.0	0.0
V7. Aquifers and High Water	0.0	0.0	0.0	0.0	0.0	0.0	0.0	0.0	0.0	0.0	0.0
V8. Relief Type and Morphometry	0.0	0.0	0.0	0.0	0.0	0.0	0.0	0.0	0.0	0.0	0.0
V9. Engineering and geological processes	0.0	0.0	0.0	0.0	0.0	0.0	0.0	0.0	0.0	0.0	0.0
V10. Mining construction technologies	0.0	0.0	0.0	0.0	0.0	0.0	0.0	0.0	0.0	0.0	0.0
I-Vll. THE VIABILITY OF THE UNDERGROUND URBAN DEVELOPMENT	0.0	0.0	−LO	−1.0	−3.0	−5.0	−6.0	−7.0	−11.0	−14.0	−17.0
V12. The level of comfort of work and rest during the construction and operation of underground structures	0.0	0.0	−LO	−2.0	−2.0	−4.0	−6.0	−7.0	−8.0	−12.0	−15.0

(continued)

Table 6.3 (continued)

Step Vertex	0.0	1.0	2.0	3.0	4.0	5.0	6.0	7.0	8.0	9.0	10.0
I-V13. DISASTERS, EXTREME AND EMERGENCY SITUATIONS	0.0	0.0	2.0	3.0	3.0	4.0	6.0	7.0	9.0	13.0	16.0
V14. Construction, operational, management risks	0.0	1.0	2.0	2.0	3.0	5.0	6.0	8.0	12.0	15.0	17.0
I-V15. ENVIRONMENTAL RISKS	0.0	1.0	1.0	1.0	1.0	1.0	1.0	1.0	1.0	1.0	1.0
1-V16. ECONOMIC RISKS	0.0	1.0	1.0	1.0	1.0	1.0	1.0	1.0	1.0	1.0	1.0
V17. Staff qualifications	0.0	0.0	0.0	−LO	−2.0	−2.0	−4.0	−6.0	−7.0	−8.0	−12.0
V18. Industrial Safety	0.0	0.0	−2.0	−5.0	−7.0	−9.0	−12.0	−17.0	−22.0	−29.0	−37.0
V19. Quality and construction time	0.0	0.0	0.0	0.0	−1.0	−2.0	−2.0	−4.0	−6.0	−7.0	−8.0

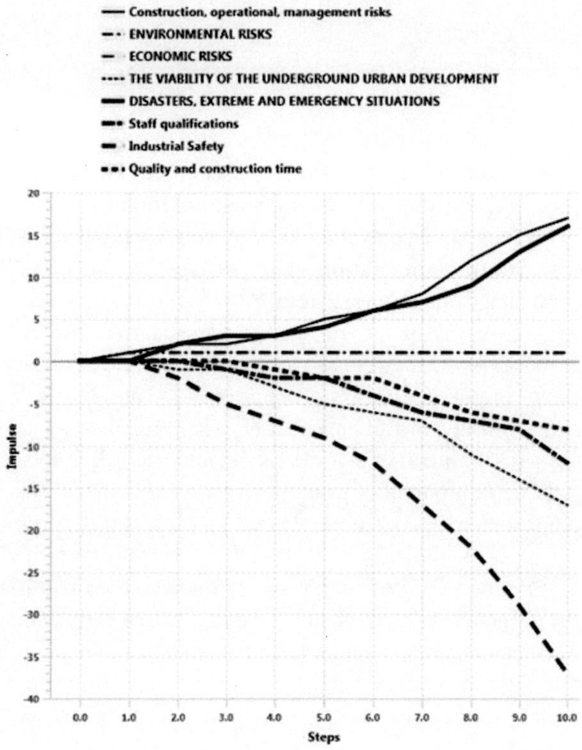

Fig. 6.12 Graphs of pulsed processes, vertices V_{14}, $I - V_{15}, I - V_{16}, I - V_{11}, I - V_{13}, V_{17}, V_{18}, V_{19}$. Scenario No. 2

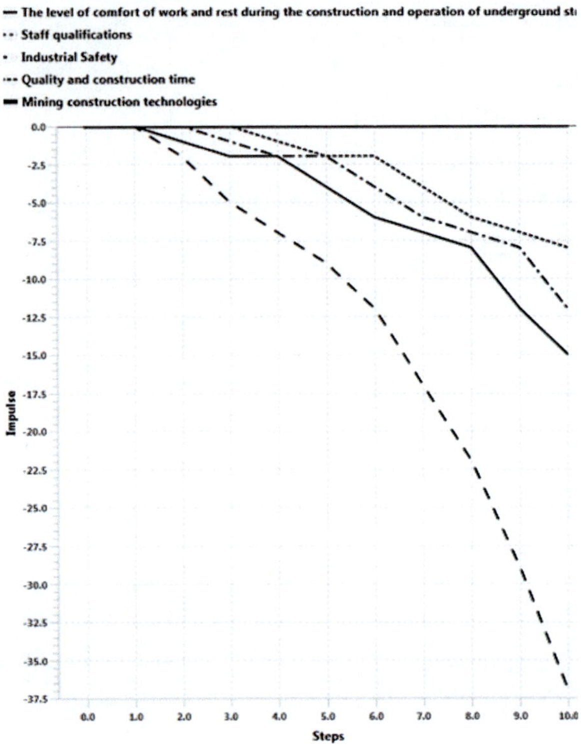

Fig. 6.13 Graphs of pulsed processes, vertices V_{12}, V_{17}, V_{18}, V_{19}, V_{10}. Scenario No. 2

As can be seen from Fig. 6.21, scenario No. 4 can be considered the best of those considered, although its results are not too different from the results of scenario No. 3 (Fig. 6.20). If you set the task of minimizing the cost of resources for the particular scenario implementation, then perhaps scenario No. 3 will be the best, with fewer control actions in the system.

A comparison of the results of scenarios No. 3 and No. 4 with the results of scenario No. 1 (Fig. 6.18), in which the control action is applied to only one vertex, shows that it is inferior to scenarios No. 3 and No. 4. So, for example, the pulse value at the vertices of Industrial safety (V_{10}) reaches 30, and according to scenario No. 3, the pulse value at this vertex is 78, and according to scenario No. 4 pulse value is 85.

If we compare the simulation results of scenario No. 2 (Fig. 6.19) with the results of other scenarios, it is obvious that without countering possible risks, the development scenarios of the natural-technical geosystem system will be extremely pessimistic.

The modeling of scenarios for the analyzed complex system is carried out under the influence of various internal and external disturbances and control impulse effects. The results of the conducted cognitive modeling make it possible to judge that the cognitive models, which systematize and structure various information about the underground construction system, correspond to the real system and can be used to anticipate the possible processes of situations in the system under the influence of

Table 6.4 The results of a computational experiment. Scenario No. 3

Step Vertex	1.0	2.0	3.0	4.0	5.0	6.0	7.0	8.0	9.0	10.0
VI. Mountain and hydrostatic pressure, seismic impact	0.0	0.0	−LO	0.0	2.0	−1.0	−1.0	11.0	−6.0	−31.0
V2. Surface Load Static Load Index	0.0	0.0	0.0	0.0	0.0	0.0	0.0	0.0	0.0	0.0
V3. The indicator of the static load of the surrounding soil massif	0.0	0.0	0.0	−1.0	−3.0	−3.0	1.0	0.0	−7.0	10.0
V4. Existing underground facilities	0.0	1.0	1.0	−1.0	−2.0	0.0	−2.0	−3.0	14.0	2.0
I-V5. GENETIC TYPE AND LITHOLOGICAL COMPOSITION OF SOILS	0.0	0.0	0.0	0.0	−LO	3.0	−3.0	1.0	0.0	−7.0
V6. Estimated soil resistance	0.0	0.0	−LO	−3.0	−2.0	4.0	3.0	−8.0	10.0	22.0
V7. Aquifers and High Water	0.0	0.0	0.0	0.0	0.0	−1.0	−3.0	−4.0	−2.0	−4.0
V8. Relief Type and Morphometry	0.0	0.0	0.0	0.0	−1.0	−3.0	−4.0	−2.0	−4.0	−9.0
V9. Engineering and geological processes	1.0	1.0	2.0	3.0	1.0	0.0	6.0	1.0	−12.0	26.0
V10. Mining construction technologies	1.0	1.0	0.0	−1.0	1.0	2.0	−3.0	3.0	15.0	−26.0
1-V11. THE VIABILITY OF THE UNDERGROUND URBAN DEVELOPMENT	0.0	2.0	1.0	3.0	3.0	7.0	8.0	13.0	19.0	50.0
V12. The level of comfoitof work and rest during the construction and operation of underground structures	0.0	0.0	4.0	3.0	5.0	5.0	9.0	10.0	16.0	23.0

(continued)

Table 6.4 (continued)

Step Vertex	1.0	2.0	3.0	4.0	5.0	6.0	7.0	8.0	9.0	10.0
1-V13. DISASTERS. EXTREME AND EMERGENCY SITUATIONS	1.0	−1.0	−5.0	−8.0	−7.0	−9.0	−15.0	−28.0	−34.0	−23.0
V14. Construction, operational, management risks	0.0	−4.0	−7.0	−7.0	−11.0	−14.0	−15.0	−18.0	−24.0	−29.0
1-V15. ENVIRONMENTAL RISKS	0.0	0.0	0.0	1.0	3.0	1.0	−8.0	−9.0	5.0	−12.0
I-VI6. ECONOMIC RISKS	0.0	−2.0	−2.0	−2.0	−2.0	−2.0	−2.0	−3.0	−4.0	−3.0
V17. Staff qualifications	1.0	1.0	1.0	5.0	4.0	6.0	6.0	10.0	11.0	17.0
V18. Industrial Safety	0.0	0.0	7.0	14.0	18.0	17.0	28.0	42.0	62.0	78.0
V19. Quality and construction time	1.0	2.0	2.0	2.0	6.0	5.0	7.0	7.0	11.0	12.0

various disturbing and controlling factors. The developed author's software system allows in the process of pulse modeling and analysis of the obtained results to introduce control or exciting actions at any stage of modeling. This allows to change (correct) scenarios in the dynamics of creating a model, to determine the effects that bring the processes closer to the desired. The developed methodology and tools made it possible to combine the assessment of the impacts and relationships of geological factors, technogenic and structural-functional types for the study of the underground objects' construction.

6.3 Study of the Plot Suitability for Urban Underground Construction

The significant changes that have occurred in recent decades in the life of large cities require a scientific understanding of the new realities and the most likely prospects for their further development. Underground urbanism, which is an integral component of modern cities, has already gone beyond the boundaries of individual local objects and is becoming a systemic factor in the existence of large cities. Foresight of future changes in the development of modern cities should be based on a reliable scientific and methodological foundation, which should ensure the development of plot and underground development as a whole.

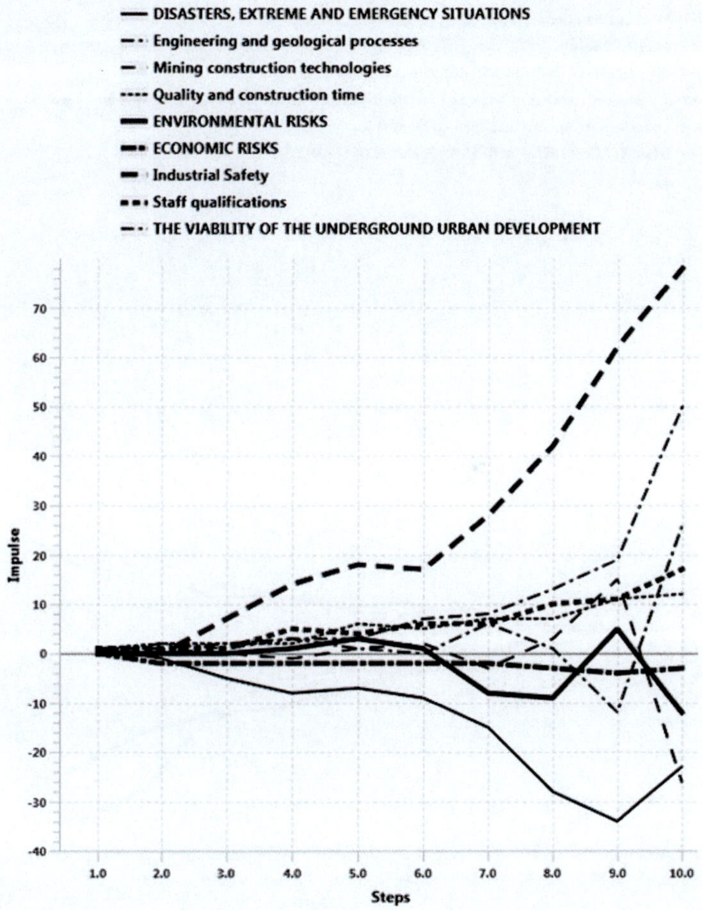

Fig. 6.14 Graphs of pulsed processes, vertices $I - V_{13}$, V_9, V_{10}, V_{19}, $I - V_{15}$, $I - V_{16}$, V_{18}, V_{17}, $I - V_{11}$. Scenario No. 3

High-tech development, including underground development of modern cities, where innovations are the base for competitiveness, as well as global ecological, economic, and demographic challenges have produced a novel world model—sustainable planetary development. The modern concept of sustainable development regarding the city planning should take into account the future needs, which implies the capability to satisfy current needs of society without causing harm to future generations (Vernadsky 2012). The important aspect of sustainable development is the potential for timely reaction to possible environmental changes, and minimization of technogenic and ecological impacts. This concept changes the common strategy for engineering projects and replaces the traditional vision of local problems with the position of systemic consideration of large natural-technical and social problems. In the concept of sustainable development of modern cities, underground urbanism

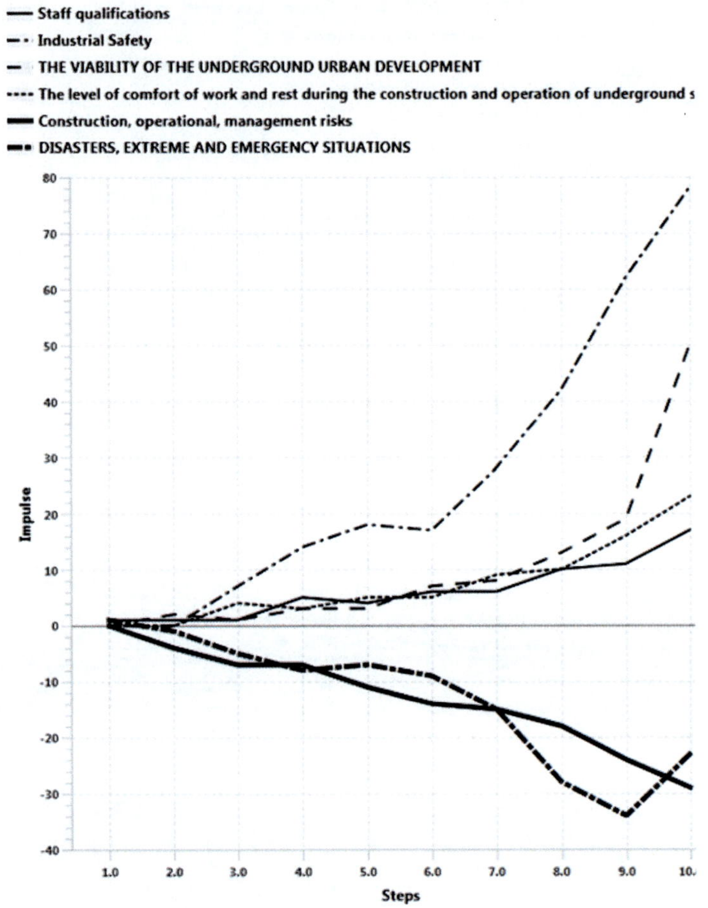

Fig. 6.15 Graphs of pulsed processes, vertices $V_{17}, V_{18}, V_{19}, I - V_{11}, V_{12}, V_{14}, I - V_{13}$. Scenario No. 3

is of particular importance, since an effectively planned underground infrastructure improves the quality of life and environmental safety to a much greater extent than a similar functional system on the surface.

The purpose of this chapter is to demonstrate the possibilities of sharing the foresight methodology for constructing of scenarios alternatives and the methodologies of cognitive and imitative modeling for creating scenarios of the underground space of urban environment development and the evaluating of underground urbanism (Pankratova and Gorelova 2020). The use of the method of morphological analysis to build alternative heterogeneous design configurations and cognitive and imitative modeling to create their scenarios can become the scientific and methodological basis for constructing a strategic master plan of the "underground city" and contribute to the

Table 6.5 The results of a computational experiment. Scenario No. 4

Step Vertex	0.0	1.0	2.0	3.0	4.0	5.0	6.0	7.0	8.0	9.0	10.0
VI. Mountain and hydrostatic pressure, seismic impact	0.0	0.0	0.0	−1.0	0.0	2,0	−1.0	−1.0	11.0	−6.0	−31.0
V2. Surface Load Static Load Index	0.0	0.0	0.0	0.0	0.0	0.0	0.0	0.0	0.0	0.0	0.0
V3. The indicator of the static load of the surrounding soil massif	0.0	0.0	0.0	0.0	−1.0	−3.0	−3.0	1.0	0.0	−7.0	10.0
V4. Existing underground facilities	0.0	0.0	1.0	1.0	−1.0	−2.0	0.0	−2.0	−3.0	14.0	2.0
I-V5. GENETIC TYPE AND LITHOLOGICAL COMPOSITION OF SOILS	0.0	0.0	0.0	0.0	0.0	−1.0	−3.0	−3.0	1.0	0.0	−7.0
V6. Estimated soil resistance	0.0	0.0	0.0	−1.0	−3.0	−2.0	4.0	3.0	−8.0	10.0	22.0
V7. Aquifers and High Water	0.0	0.0	0.0	0.0	0.0	0.0	−1.0	−3.0	−4.0	−2.0	−4.0
V8. Relief Type and Morphometry	0.0	0.0	0.0	0.0	0.0	−1.0	3.0	−4.0	2.0	−4.0	9.0
V9. Engineering and geological processes	0.0	1.0	1.0	2.0	3.0	1,0	0.0	6.0	1.0	−12.0	26.0
V10. Mining construction technologies	0.0	1.0	1.0	0.0	−1.0	1,0	2.0	−3.0	3.0	15.0	−26.0
I-Vll. THE VIABILITY OF THE UNDERGROUND URBAN DEVELOPMENT	0.0	0.0	2.0	1.0	4.0	4,0	8.0	9.0	15.0	21.0	53.0

(continued)

Table 6.5 (continued)

Step Vertex	0.0	1.0	2.0	3.0	4.0	5.0	6.0	7.0	8.0	9.0	10.0
V12. The level of comfort of work and rest during the construction and operation of underground structures	0.0	0.0	0.0	4.0	3.0	6.0	6.0	10.0	11.0	18.0	25.0
I-V13. DISASTERS, EXTREME AND EMERGENCY SITUATIONS	0.0	1.0	−2.0	−6.0	−9.0	−8.0	−10.0	−16.0	−30.0	−37.0	−26.0
V14. Construction, operational, management risks	0.0	−1.0	−5.0	−8.0	−8.0	−12.0	−15.0	−17.0	−21.0	−27.0	−32.0
I-V15. ENVIRONMENTAL RISKS	0.0	0.0	0.0	0.0	1.0	3.0	1.0	−8.0	9.0	5.0	−12.0
I-V16. ECONOMIC RISKS	0.0	0.0	−2.0	−2.0	−2.0	−2.0	−2.0	−2.0	−3.0	−4.0	−3.0
V17. Staff qualifications	0.0	1.0	1.0	1.0	5.0	4,0	7.0	7.0	11.0	12.0	19.0
V18. Industrial Safety	0.0	0.0	1.0	9.0	16.0	20.0	19.0	31.0	46.0	68.0	85.0
V19. Quality and construction time	0.0	1.0	2.0	2.0	2.0	6.0	5.0	8.0	8.0	12.0	13.0

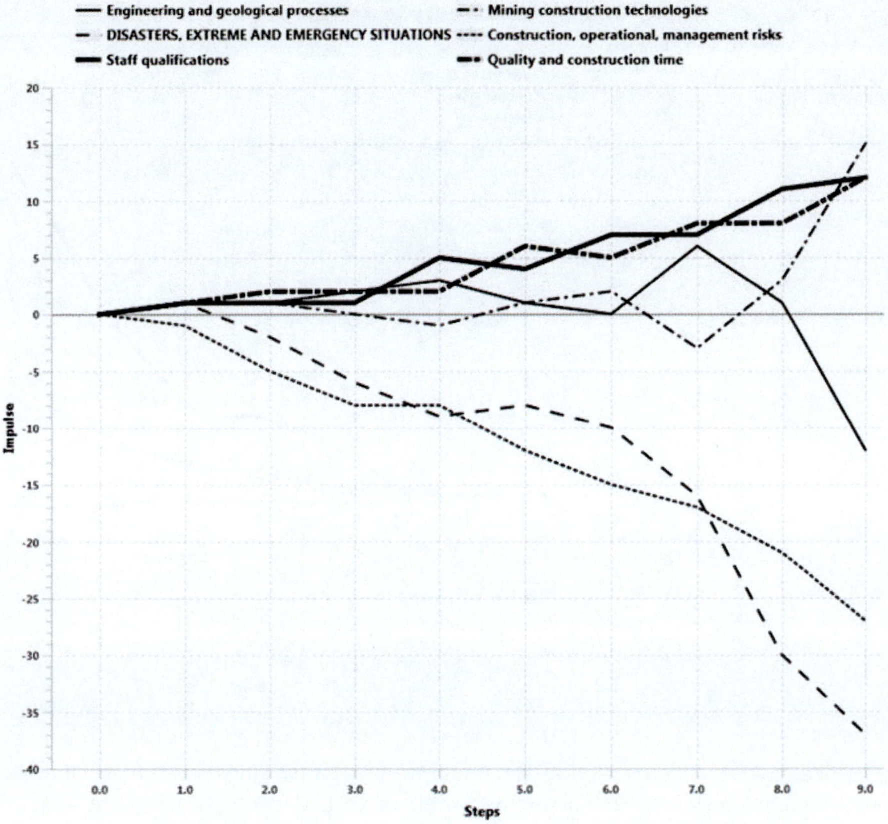

Fig. 6.16 Graphs of pulsed processes, vertices $I - V_{13}$, V_9, V_{10}, V_{17}, V_{19}. Scenario No. 4

developing urban underground space. Historically established approaches to developing underground space include resource-based, directive-based, city construction-based, and complex approaches. The complex approach is shown in Konykhov (2010) to allow the devising of the most optimal and rational system of utilizing underground space. Prediction of future changes, the capability to swap functions of underground objects, and the corresponding urban policy, planning, and management of modern city development should be based on a sound scientific-methodological base, devised to provide systemic surface and underground urban development as a whole (Haiko 2018; Pankratova et al. 2019a, 2016).

Various directions of implementing system approach for planning urban surface construction in megacities are known (Sterling et al. 2012; Kartosiya 2015; Owen et al. 2000). Analyzing the trends of the future development of the underground space of megacities are considered the individual projects of urban underground studies that characterize the directions of urban underground construction in the near and medium perspective (Gilbert, et al. 2013). So, in Chicago, the second largest economic center in the United States, it is planned to build an underground city with

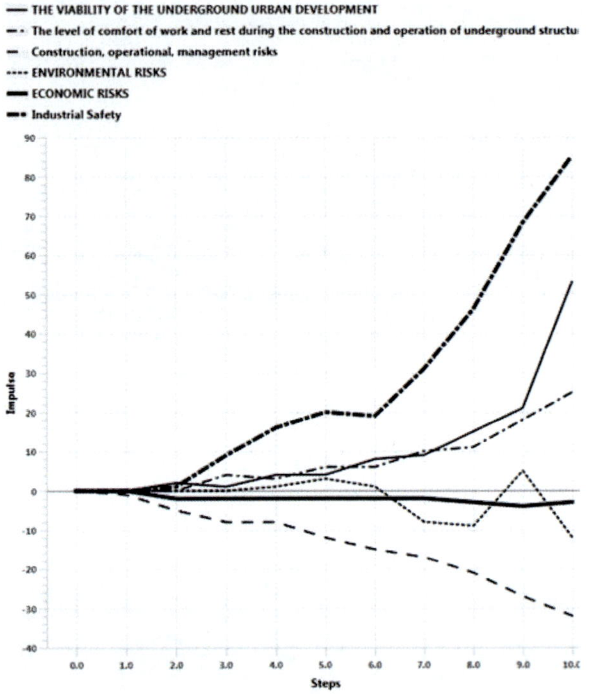

Fig. 6.17 Graphs of pulsed processes, vertices V_{12}, V_{14}, V_{15}, V_{16}, V_{18}, $I - V_{11}$. Scenario No. 4

a vertical layout, which will have 100 underground floors. The development of urban environments currently actualizes the problems of underground urbanism. One of the main among them is the problem of choosing a construction plot and assessing its suitability for many indicators.

6.3.1 Cognitive Modeling of the Plot Suitability Scenarios for Urban Underground Construction

Study on the basis of morphological analysis method on the foresight stage of two plots in Kyiv for construction of underground parking lots with different characteristics is considered. The two underground parking lot plots with different characteristics are investigated. The first construction plot is found at the Shevchenkivsky district of the Kyiv City at the Peremohy avenue, and the second is also found at the Shevchenkivsky district between the Bulvarno-Kudriavska and Honchara streets. The morphological tables (4.31–4.33) were estimated using the available data of geological and technogenic nature, and the dependencies between parameters were obtained by expert estimation (Pankratova et al. 2019a). To build a morphological table for construction plots, some sets of characteristics that affect the result in the same way, in this study, were aggregated into the following parameters: suitability of the plot,

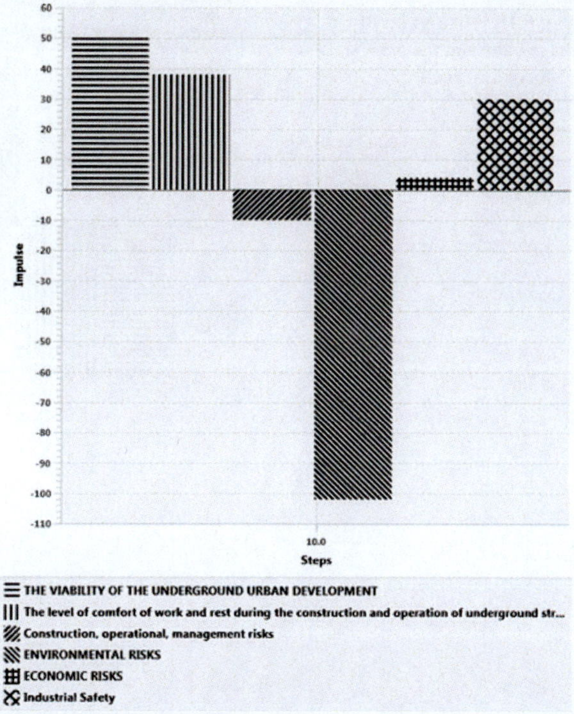

Fig. 6.18 Histogram of pulse values at the 10th step of modeling according to Scenario No. 1

the level of dynamic load, the indicator of static load from surface development, the indicator of the static load of the surrounding soil mass, the impact of existing underground facilities, genetic type and lithological composition of soils, estimated soil stable, influence of aquifers and water floods, type of terrain and morphometry, engineering and geological processes, and geotechnologies for the construction of underground structures. Studies using the MMAM showed that the second plot refers to the alternative most suitable for underground construction, as evidenced by the absolute probability indicators of the alternative suitability of the plot having values of 0,688 and 0,993, respectively (Pankratova et al. 2019a). Therefore, for cognitive modeling, the results obtained for the second plot are used.

In accordance with the cognitive modeling strategy proposed above, we will conduct all the steps sequentially.

The first stage: cognitive model development. In developing the cognitive model, the principle of sequential refinement of the model was used: initially, a cognitive map was built (action 1), then not on its basis a model of a weighted sign digraph was constructed and studied (action 2).

Action 1: *creation of cognitive map G*. Let us call the cognitive map *G* as "Underground Construction". When developing a cognitive model, it is convenient to present

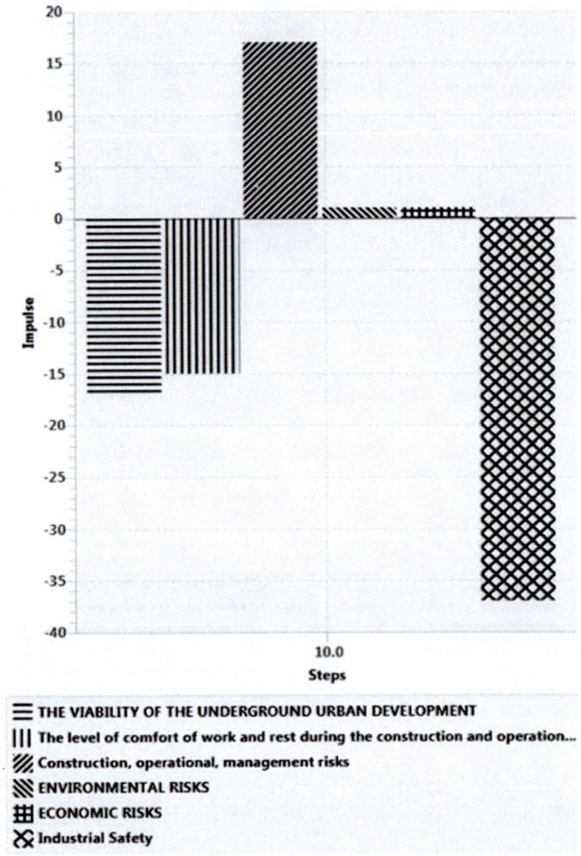

Fig. 6.19 Histogram of pulse values at the 10th step of modeling according to Scenario No. 2

the analyzed and systematized information about the vertexes and their relationships (Gorelova and Pankratova 2015) in the form of Table 6.6 and Table 6.7.

Cognitive map G is shown in Fig. 6.22. Solid lines correspond to positive arcs " +1", dotted to negative "–1".

Data grid in which you can make changes when developing and adjusting the cognitive map are shown in Fig. 6.23. The relationship matrix R_G is shown in Fig. 6.24.

***The second stage*: *analysis of the properties of the cognitive map G*.** An analysis of the properties of the cognitive map G included determining the characteristics of the graph, analyzing paths and cycles, analyzing stability to perturbations and structural stability.

(1a) *Characterization of the graph.* In Fig. 6.25, graph characteristics are given: the number of vertexes and arcs is calculated, and the number of incoming (p^+) and outgoing (p^-) arcs is determined.

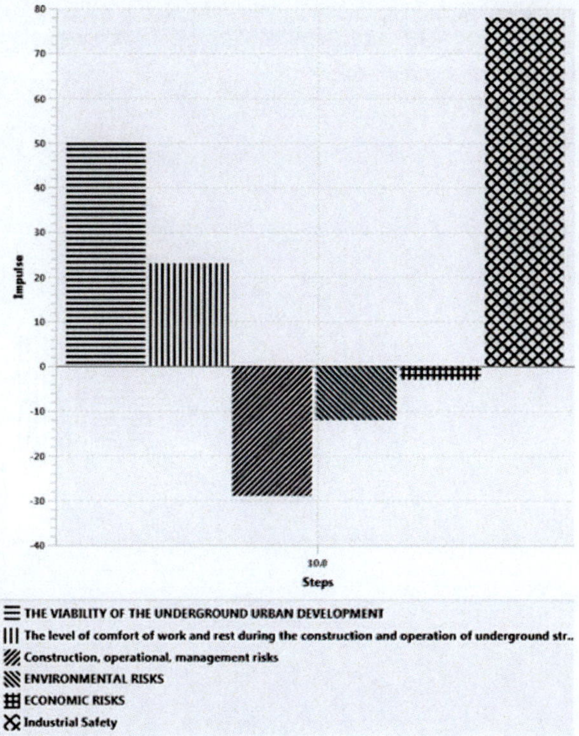

Fig. 6.20 Histogram of pulse values at the 10th step of modeling according to Scenario No. 3

(1b) *Definition and analysis of the paths for model G.* The study of different paths from top to top on a cognitive map allows us to comprehend how reasonably the cause-effect chains of the cognitive map "Underground Construction" are defined, in which the main characteristics of the plot intended for underground construction are specified. An analysis of the paths makes it possible to interpret their meaning as cause-effect chains and use the information further to support decisions being made. In addition, the way shows the variety of opportunities to achieve goals, their number may be large and far from obvious to the decision-maker. It also becomes clear which vertexes are included in each path and are "touched" on this path, which gives an occasion to evaluate its desirability or non-desirability.

Figure 6.26 shows examples of paths between the vertexes V_{10} (geotechnologies for the construction of underground structures) and V_0 (suitability of the plot); all such paths 16; a) is one of the positive (no negative arcs or an even number of negative arcs) and b) is one of the negative (odd number of negative arcs) paths.

(1c) Analysis of cycles and structural stability of the cognitive map *G* are presented in Fig. 6.27.

The analysis of the cognitive map cycles showed that this system contains 47 cycles, among them 27 cycles of negative (stabilizing) feedback. This indicates that the analyzed system is structurally stable (Roberts 1978; Kulba et al. 2002; Atkin

Fig. 6.21 Histogram of pulse values at the 10th step of modeling according to Scenario No. 4

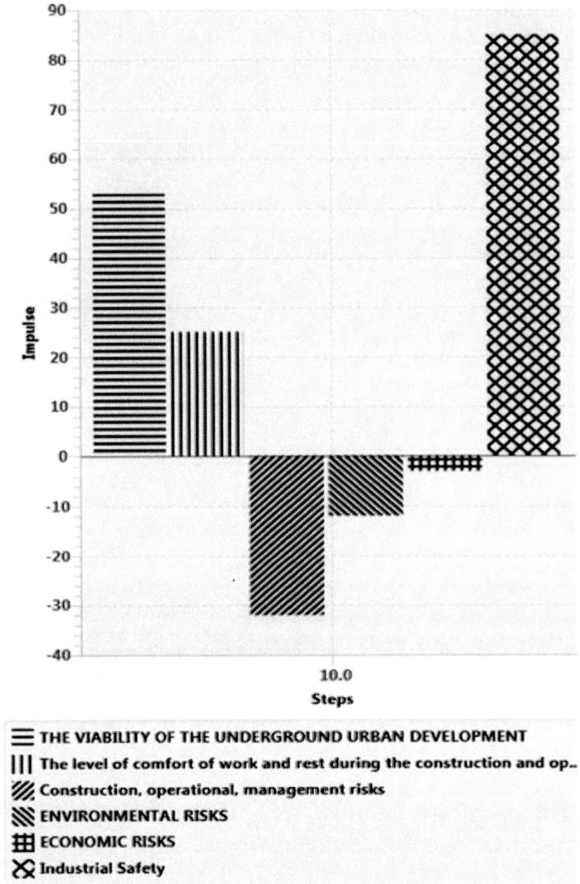

Table 6.6 Vertexes V_i of cognitive map G

Code	Name of the vertex	Vertex assignment
V_0	Suitability of the plot	Indicative, target
V_1	The level of dynamic load	Basic, disturbing
V_2	The indicator of static load from surface development	Basic, disturbing
V_3	The indicator of the static load of the surrounding soil mass	Basic, disturbing
V_4	The impact of existing underground facilities	Disturbing
V_5	Genetic type and lithological composition of soils	Basic
V_6	Estimated soil stable	Basic
V_7	Influence of aquifers and water floods	disturbing
V_8	Type of terrain and morphometry	Basic
V_9	Engineering and geological processes	Disturbing
V_{10}	Geotechnologies for the construction of underground structures	Basic, governing

Table 6.7 Fragment of a relation matrix and cognitive map *G*

Code	Vertex "cause" V_i	Vertex "effect" V_j	Sign of relationship
e_{14}	V_1 The level of dynamic load	V_4 The impact of existing underground facilities	+1
e_{19}		V_9 Engineering and geological processes	–1
e_{110}		V_{10} Geotechnologies for the construction of underground structures	+1
…			
e_{101}	V_{10} Geotechnologies for the construction of underground structures	V_1 The level of dynamic load	–1
e_{104}		V_4 The impact of existing underground facilities	+1
e_{106}		V_6 Estimated soil stable	+1
e_{100}		V_0 Suitability of the plot	+1

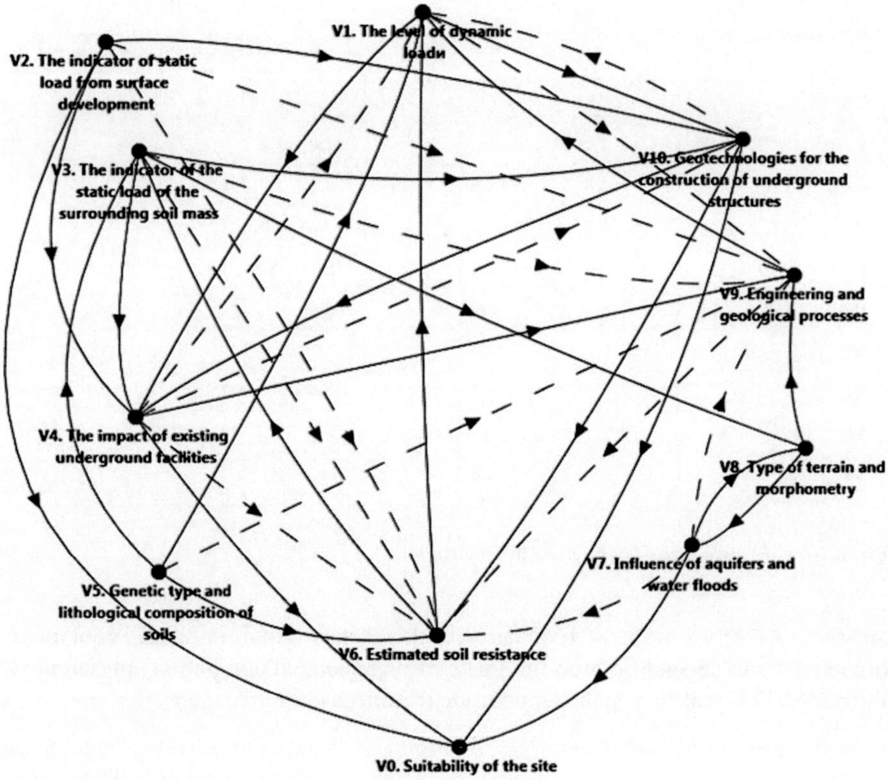

Fig. 6.22 Cognitive map G

Vertices | Edges

Full name	Short name	Weight	Impulse
The level of dynamic loadи	V1	0.0	0.0
The indicator of static load from surface development	V2	0.0	0.0
The indicator of the static load of the surrounding soil mas	V3	0.0	0.0
The impact of existing underground facilities	V4	0.0	0.0
Genetic type and lithological composition of soils	V5	0.0	0.0
Estimated soil resistance	V6	0.0	0.0
Influence of aquifers and water floods	V7	0.0	0.0
Type of terrain and morphometry	V8	0.0	0.0
Engineering and geological processes	V9	0.0	0.0
Geotechnologies for the construction of underground struc	V10	0.0	0.0
Suitability of the site	V0	0.0	0.0

Fig. 6.23 Grid Data entry

Connectivity matrix

- Infl./ Dep.	V1	V2	V3	V4	V5	V6	V7	V8	V9	V10	V0
V1	X			1.0					-1.0	1.0	
V2		X		1.0		-1.0			-1.0	1.0	1.0
V3			X	1.0		-1.0		1.0	-1.0	1.0	
V4	-1.0			X		-1.0			1.0	-1.0	1.0
V5	1.0		1.0		X	1.0			-1.0		
V6	1.0		1.0			X					
V7						-1.0	X	1.0	-1.0		1.0
V8							1.0	X	1.0		
V9	1.0					-1.0			X		

Fig. 6.24 Cognitive map G of relationship matrix R_G

and Casti 1977; Atkin 1997). Evaluation of this fact as "good" or "bad" is not unambiguous, it can depend both on the goals of management and on the "preferences" of the decision-maker, which can upgrade the initial cognitive map.

(1d) *Analysis of stability to disturbances.* Figure 6.28 shows the results of calculating the eigenvalues (roots of the characteristic equation) of the matrix RG. Their determination is necessary for the analysis of the stability of the system to perturbations and the initial value (Kulikova et al. 2005; Casti 1979)

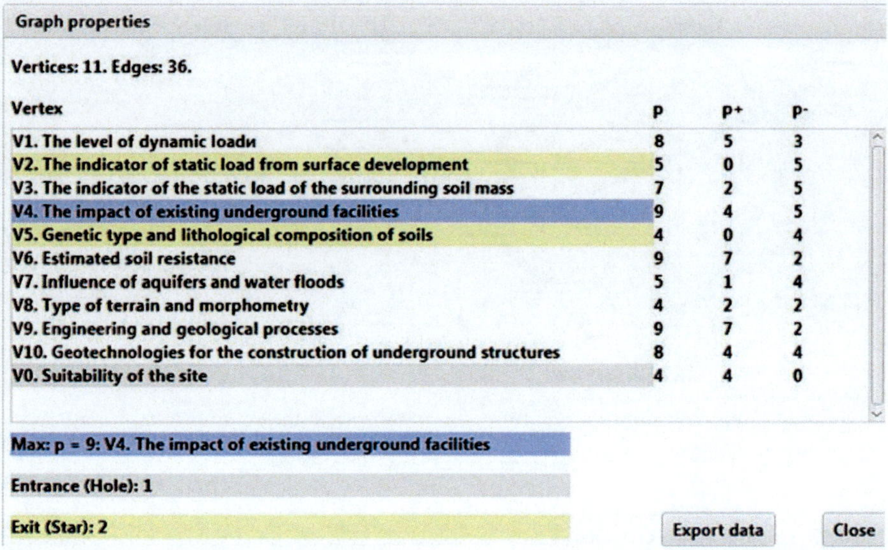

Graph properties

Vertices: 11. Edges: 36.

Vertex	p	p+	p-
V1. The level of dynamic loadи	8	5	3
V2. The indicator of static load from surface development	5	0	5
V3. The indicator of the static load of the surrounding soil mass	7	2	5
V4. The impact of existing underground facilities	9	4	5
V5. Genetic type and lithological composition of soils	4	0	4
V6. Estimated soil resistance	9	7	2
V7. Influence of aquifers and water floods	5	1	4
V8. Type of terrain and morphometry	4	2	2
V9. Engineering and geological processes	9	7	2
V10. Geotechnologies for the construction of underground structures	8	4	4
V0. Suitability of the site	4	4	0

Max: p = 9: V4. The impact of existing underground facilities

Entrance (Hole): 1

Exit (Star): 2

Export data

Close

Fig. 6.25 Characteristics of the graph G

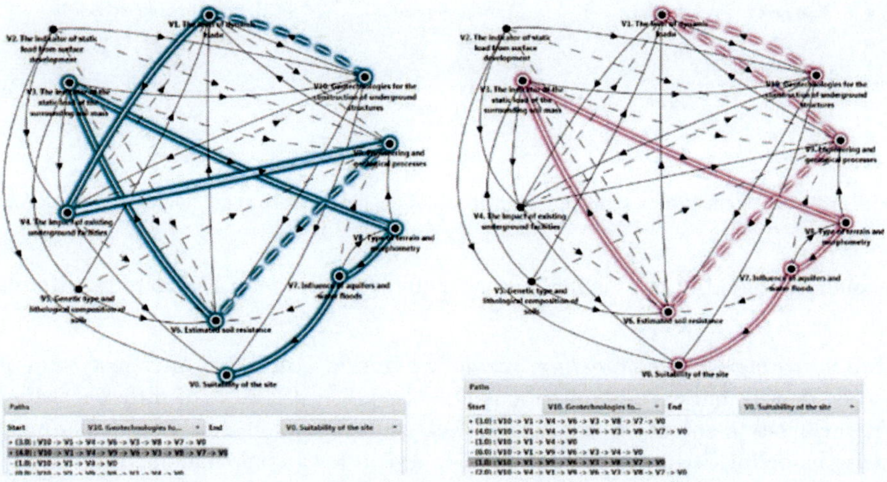

a b

Fig. 6.26 Ways of the graph G from the vertex V_{10} to the vertex V_0

The stability criterion $|M| < 1$ is used, where $|M|$ is the maximum modulus eigenvalue of the matrix R_G. Since in this case $|M| = 1.9309 > 1$, the system G is not stable either to perturbation or to the initial value. This indicates that the slightest deviations at the vertexes remove the system from a stable state. Making a decision

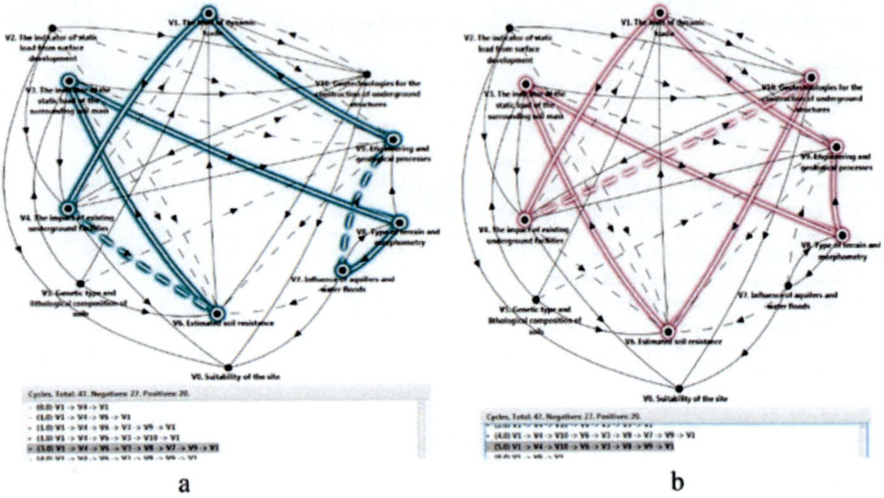

Fig. 6.27 Cognitive map *G* cycles

Eigenvalues

#	Real part	Imaginary part	Module (1.9309)
0	-0.6809	1.9309	1.9309
1	-0.6809	-1.9309	1.9309
2	0.7289	1.2541	1.2541
3	0.7289	-1.2541	1.2541
4	-0.7276	0.0	0.7276
5	-0.1169	0.5886	0.5886

Fig. 6.28 Fragment of the calculation of the roots of the characteristic equation of the matrix R_G

about how "good"/ "bad" is in this case requires further analysis and refinement of the model.

The third stage. Pulse modeling, scenario analysis. Using the software system, it is possible to introduce perturbations $Q = \{q_i\}$, $i = 1,2, \ldots k$ of different values (normalized) to any of the vertexes, as well as their combination. Before starting a pulse modeling, it is necessary to develop a plan for a computational experiment. When introducing disturbances to the vertexes, an answer is sought to the question: "What will happen if…?". In the process of pulse modeling, it is possible to introduce disturbing influences at any modeling step. This makes it possible to change (correct) scenarios in model dynamics, to determine the effects that bring processes closer to the desired ones. Table 6.8 shows a fragment of an experimental design containing 12 scenarios.

Scenario No. 1. Let the Geotechnologies for the construction of underground structures be improved, the impulse $q_{10} = +1$ is introduced to the vertex V_{10}, the disturbance vector $Q_1 = \{q_1 = 0; q_2 = 0; \ldots q_{10} = +1\}$.

Table 6.8 Fragment of the experimental design for model *G*

No	The vector of disturbances $Q = \{\ldots\}$	No. of vertexes V_i to which momenta are introduced									
		1	2	3	4	5	6	7	8	9	10
One vertex perturbation											
1	**Scenario No. 1**: $q_{10} = +1$; $Q_1 = \{q_1 = 0; q_2 = 0; \ldots q_{10} = +1\}$										+1
…											
Four vertexes perturbation											
12	**Scenario No. 2**: $q_2 = -1$; $q_4 = -1$, $q_9 = +1$, $q_{10} = +1$, $Q_2 = \{0; q_2 = -1; \ldots q_4 = -1; \ldots q_9 = +1; q_{10} = +1\}$		–1		–1					+1	+1

Table 6.9 presents the results of calculations of impulse processes at the vertexes at 10 modeling steps, the number of which in this case can be considered sufficient, because trends in situations in the system have already manifested. Figure 6.29 shows graphs of pulsed processes at the vertexes constructed according to the data in Table 6.9. The graphs are represented by two figures (a and b) with fewer vertexes in each in order to facilitate their visual analysis.

Modeling of scenario No. 1 showed that in the case of a system model presented in the form of a cognitive map *G*, the processes in it will develop in an increasing oscillatory mode with an expected change at the top of V_{10}.

Scenario No. 2. Let four vertexes be excited: there is a negative effect on the system of static load from surface development ($q_2 = -1$) and existing underground facilities ($q_4 = -1$), but geotechnical processes and geotechnologies for the construction of underground structures are assumed to be positive ($q_9 = +1$ and $q_{10} = +1$); perturbation vector $Q_2 = \{0; \ldots q_2 = -1; \ldots q_4 = -1; \ldots q_9 = +1; q_{10} = +1\}$.

The calculation results are presented in Fig. 6.30.

Analyzing the results of scenario modeling No. 2, we see that the same nature of the growing oscillatory processes is observed as in scenario No. 1. Model *G* turns out to be impulsively unstable, which also follows from previous results of the cognitive map *G* study for stability to perturbations. But model *G* is structurally stable, and its instability to perturbations can be caused by the fact that all relations a_{ij} are the same, equal to unity. This may not be true.

In continuation of the research, an expert survey was conducted, with the help of which the weight coefficients of the arcs were established.

Table 6.9 Pulse processes according to scenario No. 1

Step Vertex	0.0	1.0	2.0	3.0	4.0	5.0	6.0	7.0	8.0	9.0	10.0
VI. The level of dynamic loadn	0.0	0.0	−1.0	−1.0	3.0	−4.0	−10.0	32.0	4.0	−138.0	159.0
V2. The indicator of static load from surface development	0.0	0.0	0.0	0.0	0.0	0.0	0.0	0.0	0.0	0.0	0.0
V3. The indicator of the static load of the surrounding soil mass	0.0	0.0	0.0	1.0	0.0	−4.0	2.0	7.0	−21.0	9.0	96.0
V4. The impact of existing underground facilities	0.0	0.0	1.0	0.0	−1.0	4.0	−3.0	−19.0	35.0	42.0	−180.0
V5. Genetic type and lithological composition of soils	0.0	0.0	0.0	0.0	0.0	0.0	0.0	0.0	0.0	0.0	0.0
V6. Estimated soil resistance	0.0	0.0	1.0	0.0	−4.0	2.0	7.0	−21.0	9.0	96.0	−160.0
V7. influence of aquifers and water floods	0.0	0.0	0.0	0.0	0.0	1.0	0.0	−3.0	2.0	4.0	−19.0
VS. Type of terrain and morphometry	0.0	0.0	0.0	0.0	1.0	0.0	−3.0	2.0	4.0	−19.0	13.0
V9. Engineering and geological processes	0.0	0.0	0.0	2.0	0.0	−3.0	11.0	2.0	−53.0	54.0	148.0
V10. Geotechnologiesfor the construction of underground structures	0.0	1.0	1.0	−1.0	1.0	5.0	−11.0	−4.0	59.0	−51.0	−170.0

***Action 2**. Development of a cognitive model* of the type of a weighted directed graph Ψ (6.3), (6.4). Table 6.10 shows the model data with weights. The corresponding cognitive model Ψ corresponding to Table 6.10 is shown in Fig. 6.31.

In the course of the study, an analysis was made of the stability of the model to perturbations and according to the initial value.

Analysis of stability to disturbances. The results of studying the stability of the model to perturbations are shown in Fig. 6.32. In this case, $|M| = 0.92 < 1$; therefore, the model Ψ is stable to perturbations also with respect to the initial values.

Since the structural properties of the models Ψ and G are the same, we pass to pulse modeling. During the study, 17 scenarios were analyzed: 7, it is when introducing

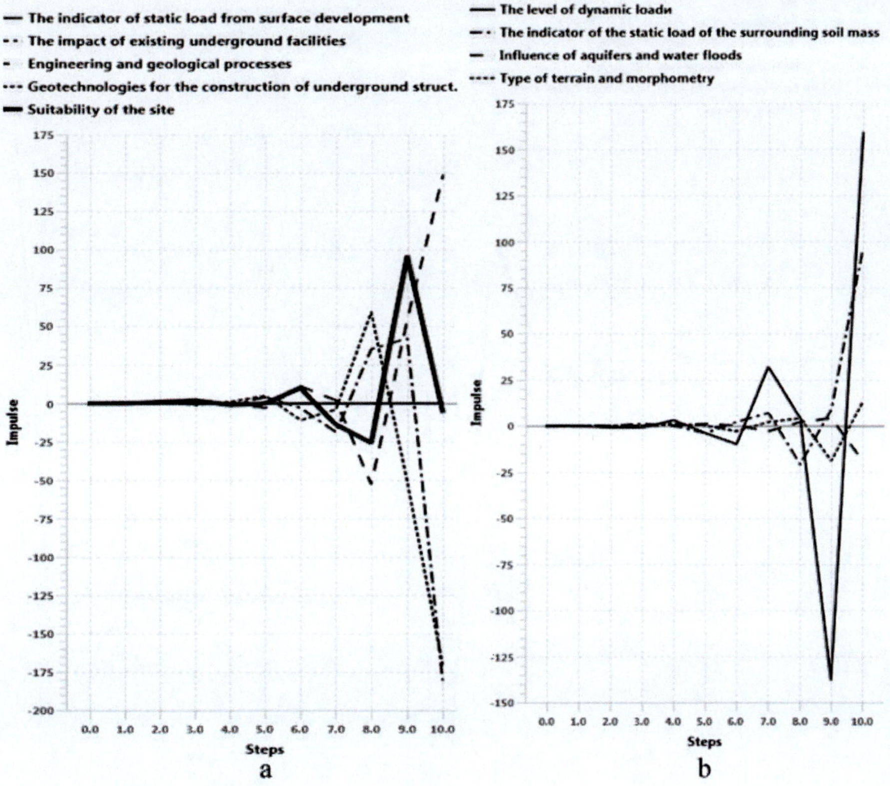

Fig. 6.29 Graphs of impulse processes according to scenario No. 1

single actions in turn into all 7 disturbing vertexes (see Table 6.6), 5, it is when introducing single actions into combinations of 2 vertexes, when introducing single actions into combinations of 3 vertexes and three scenarios when introducing single actions into combinations of 4 vertexes.

As an illustration of the results, we present the data of pulsed modeling of three scenarios (call them scenarios No. 3, No. 4, No. 5), which allow us to further suggest the main conclusions.

Scenario No. 3. Suppose one of the possible "worst" options, when the level of dynamic load in the system begins to "negatively" affect the entire system (q1 = –1), the perturbation vector is $Q4 = \{q_1 = -1; q_2 = 0; \ldots\ q_{10} = 0\}$.

The modeling results are presented in Fig. 6.33.

As can be seen from the modeling results of scenario No. 3, the negative effect of the dynamic load generates in the system a trend to deterioration in all vertexes, which is especially noticeable in the negative trend at vertex V_0 "Plot suitability".

The results of the modeling for scenarios No. 3 and No. 4, repeating the conditions of scenarios No. 1 and No. 2, are presented in Figs. 6.34 and 6.35.

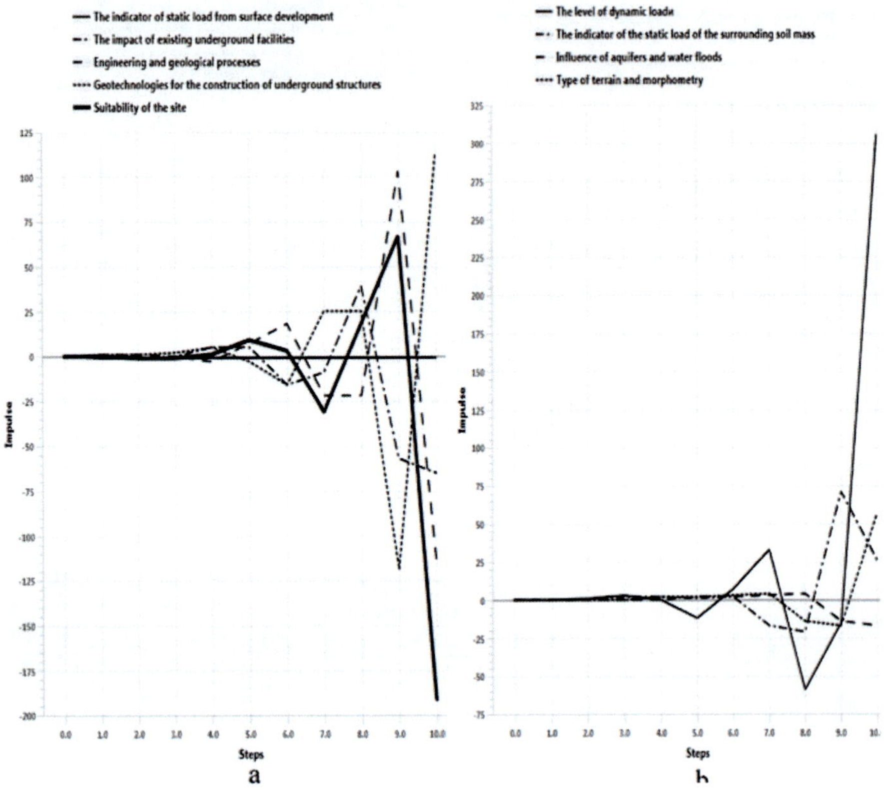

Fig. 6.30 Graphs of impulse processes according to scenario No. 2

Scenario No. 5: Let there be a negative effect on the system of static load from surface development ($q_2 = -1$) and existing underground facilities ($q_4 = -1$), but geotechnical processes and geotechnologies for the construction of underground structures are assumed to be positive ($q_9 = +1$ and $q_{10} = +1$); perturbation vector $Q_5 = \{0; \ldots q_2 = -1; \ldots q_4 = -1; \ldots q_9 = +1; q_{10} = +1\}$.

The modeling results are presented in Fig. 6.35.

Analysis of the scenarios 4 and 5 results showed the ability to withstand the negative impacts of the vertexes V_2 (The level of dynamic load) and V_4 (The impact of existing underground facilities) by positive control actions on the vertexes V_9 (Engineering and geological processes) and V_{10} (Geotechnologies for the construction of underground structures).

Scenario analysis on model Ψ showed that when disturbances are introduced into any of its vertexes, impulse processes are generated that after a certain time stabilize at a certain level; the system is momentarily stable.

Table 6.10 Weighting factors of the model Ψ

Code	Name of the vertex	V_0	V_1	V_2	V_3	V_4	V_5	V_6	V_7	V_8	V_9	V_{10}
V_0	Suitability of the plot											
V_1	The level of dynamic load					0, 6					–0, 7	0, 4
V_2	The indicator of static load from surface development	+1				0, 2		–0, 3			–0, 6	0, 7
V_3	The indicator of the static load of the surrounding soil mass					0, 1		–0, 5		0, 4	–0, 6	0, 7
V_4	The impact of existing underground facilities	+1	–0, 2					–0, 5			0, 6	–0, 7
V_5	Genetic type and lithological composition of soils		0, 4		0, 5			0, 3			–0, 2	
V_6	Estimated soil stable		0, 2		0, 7							
V_7	Influence of aquifers and water floods	+1						–0, 4		0, 6	–0, 6	
V_8	Type of terrain and morphometry								0, 8		0, 2	
V_9	Engineering and geological processes		0, 7					–0, 3				
V_{10}	Geotechnologies for the construction of underground structures	+1	–0, 7			0, 9		0, 5				

The modeling of scenarios for possible processes of the event development in the analyzed complex system is carried out under the influence of various internal and external disturbances and control impulse effects. The results of the conducted cognitive modeling make it possible to judge that the cognitive models G and Ψ, which systematize and structure various information about the underground construction system, correspond to the real system and can be used to anticipate the possible processes of situations in the system under the influence of various disturbing and controlling factors. The developed methodology and tools made it possible to combine the assessment of the impacts and relationships of geological factors, technogenic and structural-functional types for the analysis of the plot suitability of underground construction.

But when solving construction problems on a specific plot, it is necessary to introduce quantitative characteristics of vertexes into the research scheme; it may also be necessary to increase the number of vertexes and adjust the structure of the cognitive model.

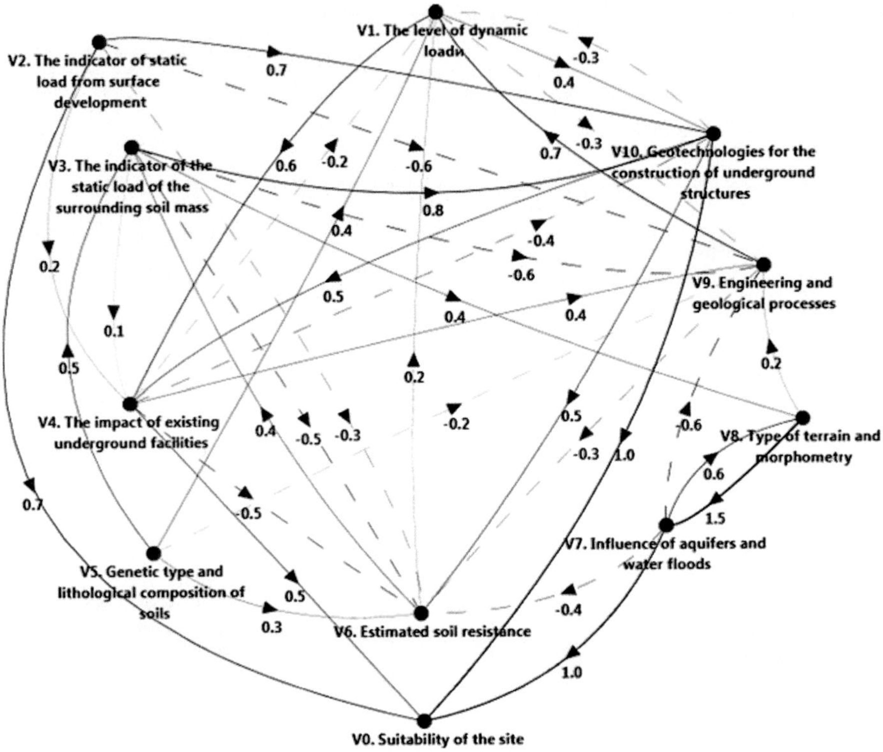

Fig. 6.31 Cognitive model Ψ

Собственные числа			
#	Действительная часть	Комплексная часть	Модуль (0.9282)
0	-0.8936	0.0	0.8936
1	-0.3152	0.8123	0.8123
2	-0.3152	-0.8123	0.8123
3	0.9282	0.0	0.9282
4	[illegible]	[illegible]	[illegible]

Fig. 6.32 Fragment of the calculation of the roots of the characteristic equation of the matrix R_Ψ

6.4 Underwater Communications Modeling Study

The space of urban environments, created by man in the process of underground construction, becomes a new, underground habitat, which should be comfortable and safe for humans. One of the most difficult problems is underground water-sewer communications that support both surface and underground urban planning. Significant advantages of water bodies' underground crossings and coastal underground infrastructure put on the agenda large-scale agenda of underground construction in

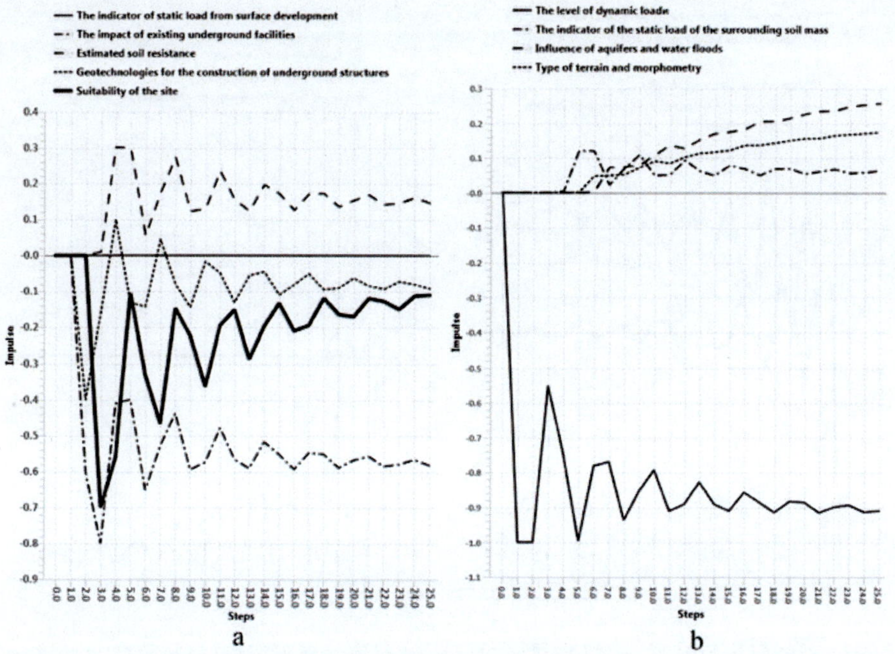

Fig. 6.33 Graphs of impulse processes according to scenario No. 3

the area of water bodies' influence (Sakellariou 2020; Kairong 2017). Here the issue of the construction of pipe and tunnel underwater duikers and justifies the priority of their creation is examined.

Solving this problem is implemented on the basis of the developed strategy of underground construction object planning based on the foresight and cognitive modeling methodologies. The suggested strategy is applied to the study of underwater communications in order to select reasonable scenarios for justification of their creation priority. In order to substantiate the choice of a tunnel duiker in comparison with a pipe duiker, the impulse modeling is carried out that allows to investigate of the process dynamics (Pankratova et al. 2022).

6.4.1 Modeling of Underwater Communications

Data on the vertices (concepts) of the hierarchical cognitive model are presented without reference to a specific territory in a generalized form in Table 6.11. We have used generalizing concepts (indicators, factors), independent of the specifics, which can be disclosed and taken into account in the future when developing the real object.

First step. Cognitive Model Development. Based on the expert and statistical analysis of the duiker system problems the following models have been developed:

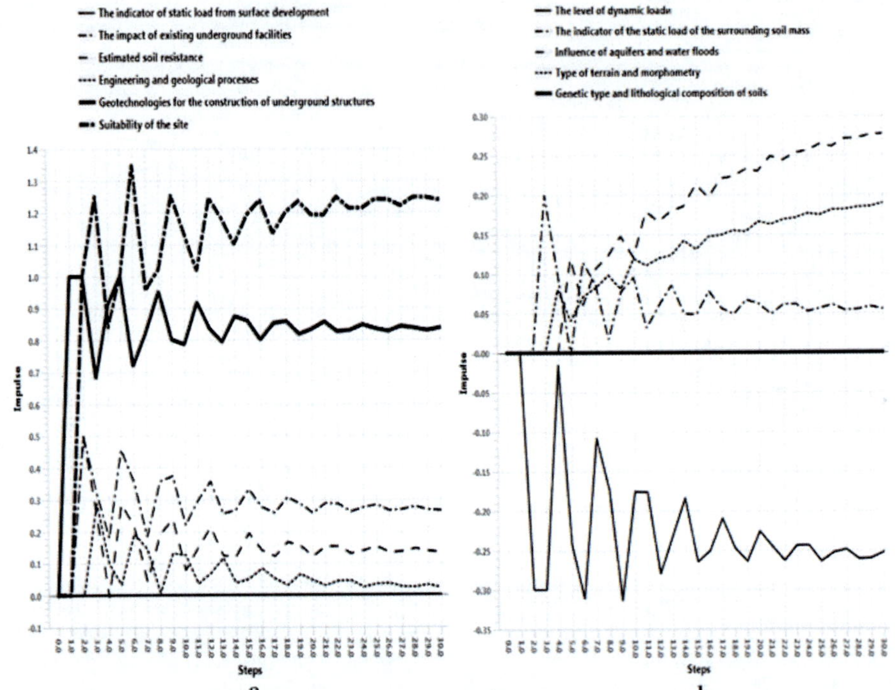

Fig. 6.34 Graphs of impulse processes according to scenario No. 4

cognitive map G1 «Pipe duiker» and G2 «Tunnel duiker». The purpose of these two models' development is to compare the options of pipe and tunnel underwater duikers' construction and justification of the priority of their creation.

When developing a cognitive model, it is convenient to present the analyzed and systematized information obtained with the method of morphological analysis at the stage of foresight in the form of cognitive map vertices. Table 6.11 shows the vertices (concepts) of the model of the cognitive map for the system "Pipe duiker".

The data about the system, grouped in Table 6.11, is visualized in the form of a cognitive map G1—model 1 (Fig. 6.36).

The solid lines of arcs in Fig. 6.36 mean that with an increase (or decrease) in the signal at the vertex V_i, the same changes occur at the vertex V_j—an increase (or decrease). The dashed lines of arcs in Fig. 6.36 mean: an increase (or a decrease) in the pulse at the vertex V_i leads to a decrease (or increase) in the pulse at the vertex V_j.

The cognitive map G1 corresponds to its connectivity matrix R_{G1} (Gorelova and Pankratova 2018). Various operations with the R_G matrix make it possible to investigate the multifaceted properties of cognitive maps. This is necessary both to verify that model G does not contradict a real complex system, and to use a cognitive map as a structure on which various scenarios for the development of situations in a real system are simulated.

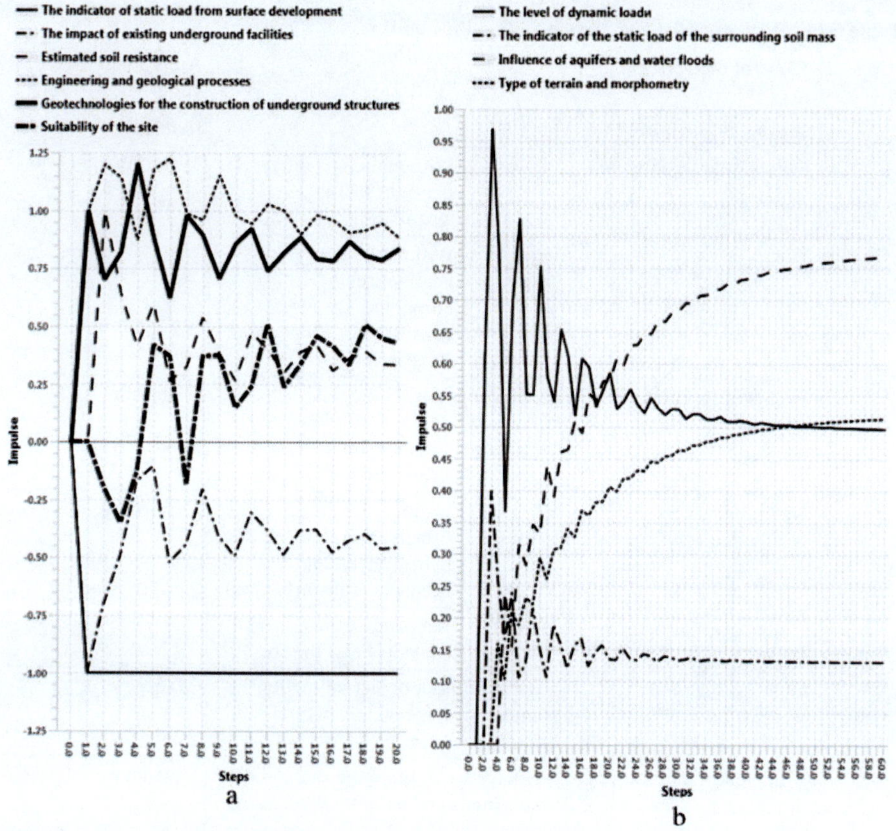

Fig. 6.35 Graphs of impulse processes according to scenario No. 5

The cognitive model is a simulation model that makes it possible not carry out an experiment on a "living" system, but to simulate its behavior and possible future development under the influence of various factors, generating new knowledge about the system. This allows to justify management decisions in certain situation.

The second step of modeling. The second step of modeling analyzes the various properties of the model is realized before using the cognitive model to determine its possible behavior. In this case, the stability properties of the model must be analyzed.

Determination of the degrees of vertices. An analysis of the degree of vertices is necessary to identify the vertices with the highest and lowest degree and determine their significance for the entire system. Figure 6.37 shows the results of determining the number (degree P) of all arcs, as well as incoming P + and outgoing P– arcs for each vertex.

Table 6.11 Vertices of the cognitive map G_1 "Pipe duiker"

Code	Name of the vertex			Assignment of the vertex
V1	System "Pipe duikers"			Indicative
V2	Events of anthropogenic origin	V2–1	With malicious intent: fighting, sabotage, terrorism	Disturbing
		V2–2	Without malice: errors, negligence, non-compliance with construction or operation standards	
V3	Events of technogenic origin	Technical and technological (technical failures, breakdowns, destruction due to technological factors, corrosion, etc.)		Disturbing
V4	Natural disasters, weather cataclysms	V4–1	Natural disasters, weather cataclysms (atmospheric, hydrosphere, and lithosphere disturbances)	Disturbing
		V4–1.1	Landslides, dips, subsidence of the soil	
V5	Object protection	Organization of object protection from anthropogenic and natural hazards		Manager
V6	Emergency condition of duikers	Abrasive wear, breakage due to holes under pipes and external stresses of worn material		Regulatory
V7	The scale of the impact of the adverse event	V7–1	Structural, or a separate section	Regulatory
		V7–2	Functional element of an object, or several sections	
		V7–3	The object is completely destroyed	
		V7–4	City area and more	
V8	Ability to function	V8–1	The object can perform all the functions	Regulatory
		V8–2	The object stops working	
V9	Resumption time	How long will it take for the object to recover from an adverse event		Regulatory
V10	Environmental consequences	Potential risks to the environmental situation		Regulatory
V11	Economic Consequences	Expected Economic Consequences in the Event of an Adverse Event		Regulatory

(continued)

Table 6.11 (continued)

Code	Name of the vertex			Assignment of the vertex
V12	Consequences for life	V12–1	Approximate number of persons whose living conditions may be violated in the event of an adverse event	Regulatory
V13	Economic resources	Expected economic losses in the event of an adverse event		Manager
V14	Organizational, technical, etc. resources	Expected resources in the event of an adverse event		Manager
V15	Investor	Additional financing for the repair of pipe siphons		Basic
V16	Integrity of the destruction system	How badly the object was damaged as a result of the impact of the undesirable event		Regulatory
V17	Material damage (minimum wages)	Expected number of losses in the event of an adverse event at the facility		Regulatory

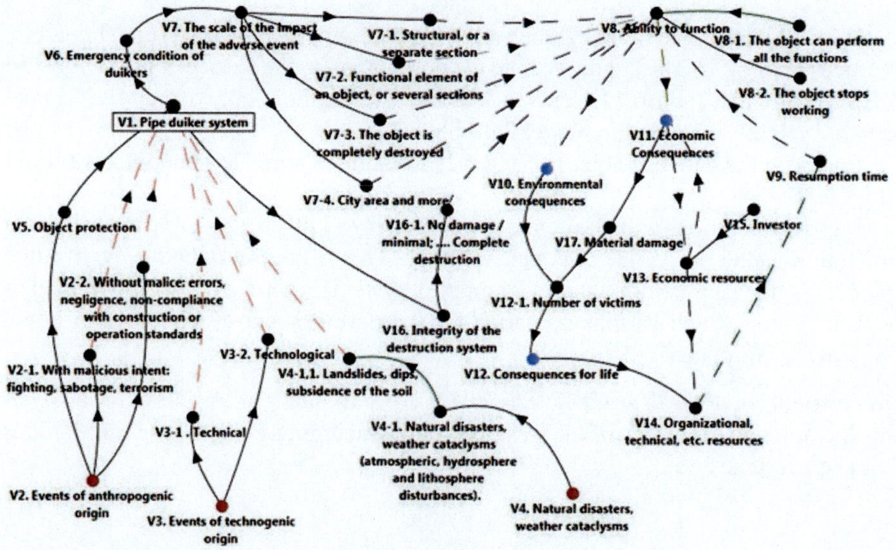

Fig. 6.36 Cognitive map G1 "Pipe duiker system"

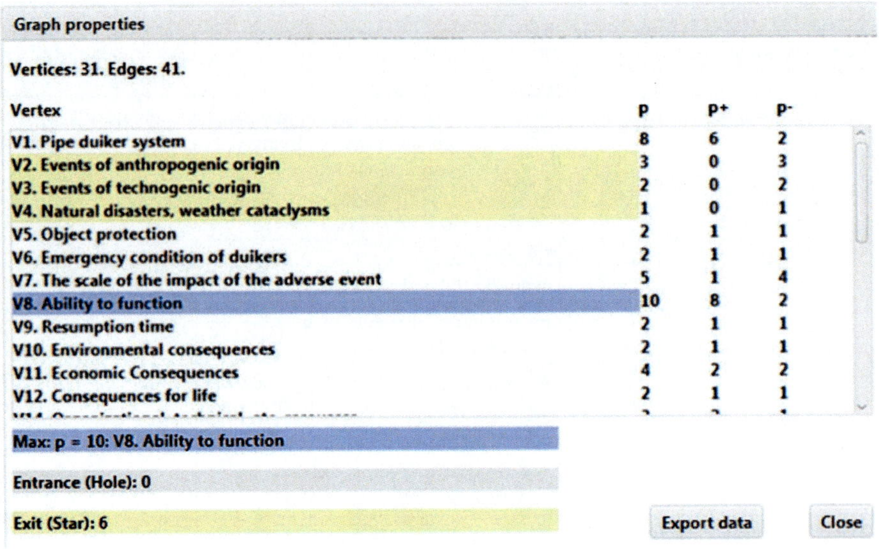

Vertex	P	P+	P-
V1. Pipe duiker system	8	6	2
V2. Events of anthropogenic origin	3	0	3
V3. Events of technogenic origin	2	0	2
V4. Natural disasters, weather cataclysms	1	0	1
V5. Object protection	2	1	1
V6. Emergency condition of duikers	2	1	1
V7. The scale of the impact of the adverse event	5	1	4
V8. Ability to function	10	8	2
V9. Resumption time	2	1	1
V10. Environmental consequences	2	1	1
V11. Economic Consequences	4	2	2
V12. Consequences for life	2	1	1

Fig. 6.37 The degrees of the vertices of the cognitive map G1

Now let us analyze model 2 connected with cognitive map G2 "Tunnel duiker". Vertices V2–V17 of the cognitive map G2 correspond to the vertices of the cognitive map G1 shown in Table 6.11. Vertices V18 and V19 for the cognitive map G2 "Tunnel duiker system" are added to model 2 (Table 6.12).

Based on the data of Tables 6.1 and 6.2, a cognitive map G2—model 2 was built (Fig. 6.38).

Various operations with the connectivity matrix R_{G_2} make it possible to investigate the multifaceted properties of cognitive maps. This is necessary both to verify that model G_2 does not contradict a real complex system, and to use a cognitive map as a structure on which various scenarios for the development of situations in a real system are simulated.

Determination of the degrees of vertices. Figure 6.39 shows the results of the analysis of the number (degree P) of all arcs, as well as incoming P+ and outgoing P– arcs for each vertex.

Table 6.12 Vertices V18, V19 of the cognitive map G2 "Tunnel duiker system"

Code	Name of the vertex	Assignment of the vertex
V1	Tunnel duiker system	Indicative
V18	Geotechnology of construction	Basic

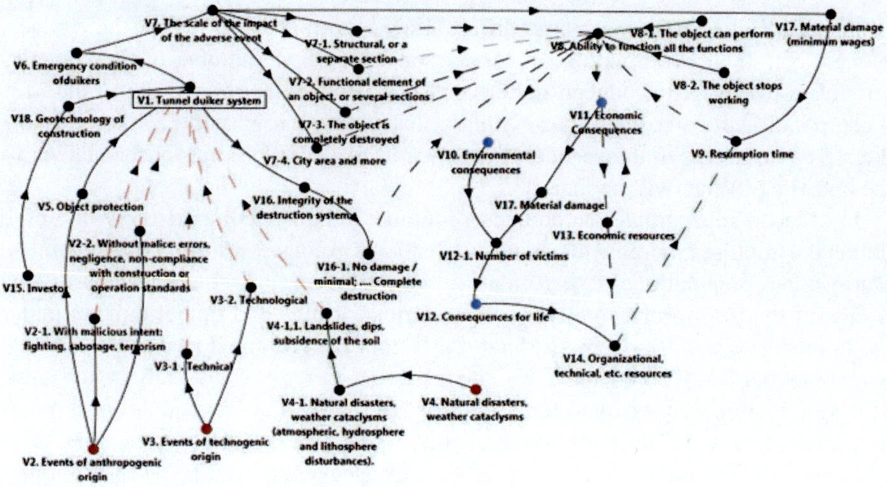

Fig. 6.38 Cognitive map G2 "Tunnel duiker system"

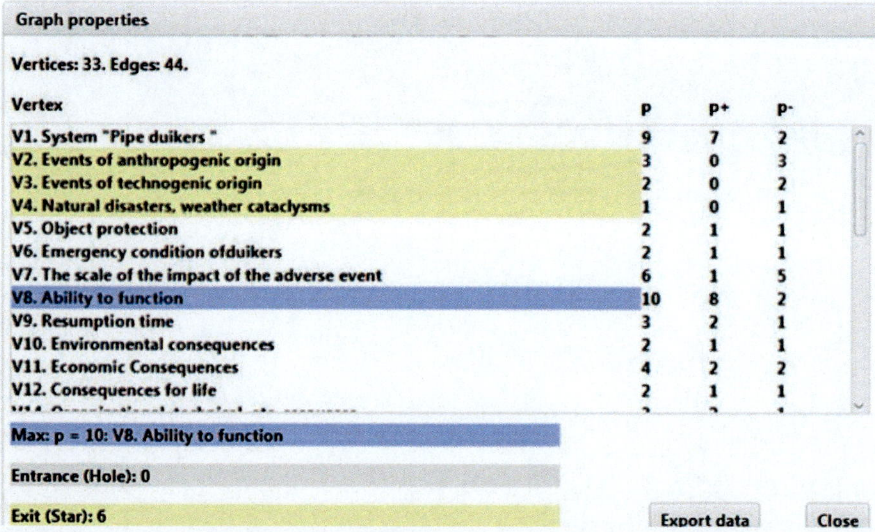

Graph properties

Vertices: 33. Edges: 44.

Vertex	P	P+	P-
V1. System "Pipe duikers"	9	7	2
V2. Events of anthropogenic origin	3	0	3
V3. Events of technogenic origin	2	0	2
V4. Natural disasters, weather cataclysms	1	0	1
V5. Object protection	2	1	1
V6. Emergency condition ofduikers	2	1	1
V7. The scale of the impact of the adverse event	6	1	5
V8. Ability to function	10	8	2
V9. Resumption time	3	2	1
V10. Environmental consequences	2	1	1
V11. Economic Consequences	4	2	2
V12. Consequences for life	2	1	1

Max: p = 10: V8. Ability to function

Entrance (Hole): 0

Exit (Star): 6

Export data Close

Fig. 6.39 Fragment of the analysis of the degree of vertices in cognitive map G2

The third step of modeling. Scenario analysis is designed to anticipate possible trends in the development of situations on the model. To generate scenarios of the development of the system, impacts are introduced into the vertices of the cognitive map in the form of a set of impulses. The impulse process formula has the form (6.6).

Let us introduce perturbations of different sizes (normalized) to any of the vertices, as well as to their combination. In connection with a large number of theoretically possible variants of introduced disturbances, it is expedient to develop a plan for a computational experiment eliminating at least almost impossible variants. Introducing disturbances to the vertices, the decision-maker is looking for the answer to the question: "What will happen if …?".

In order to substantiate the choice of a tunnel duiker in comparison with a pipe duiker the impulse modeling is carried out. It allows to consider the process dynamics. During the computational experiment, numerous scenarios were considered. The results of the most representative four scenarios for pipe and tunnel duikers under the influence of anthropogenic and natural factors are presented in Figs. 6.40, 6.41, 6.42, 6.43, 6.44, 6.45, 6.46 and 6.47. The results (see Figs. 6.40, 6.42, 6.44 and 6.46) of a computational experiment for Scenarios No. 1–No. 4 at 10 simulation steps are presented in the form of distributions: Ability to function V8, Environmental consequences V10, Economic Consequences V11, Consequences for life V12, With malicious intent (fighting, sabotage, terrorism) V2–1 and Natural disasters (Landslides, dips, subsidence of the soil) V4–1. Their representation in the form of histograms is shown in Figs. 6.41, 6.43, 6.45 and 6.47.

The following scenarios were investigated.

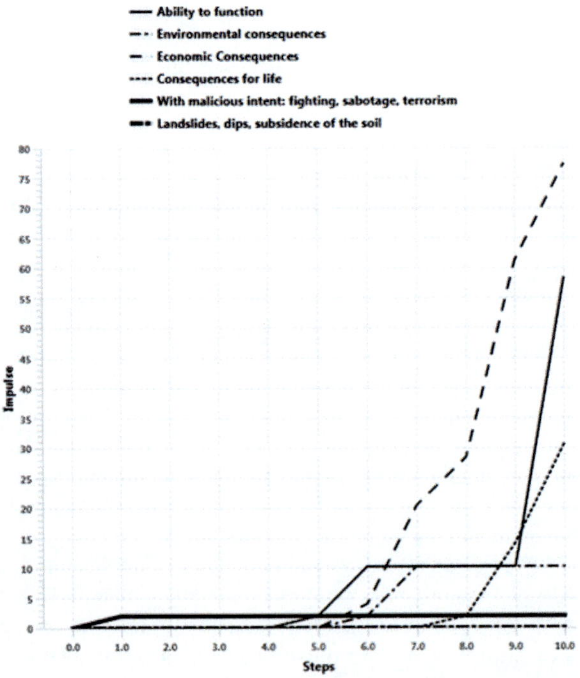

Fig. 6.40 Graphs of pulsed processes, from the first to the tenth step of modeling. Scenario No. 1

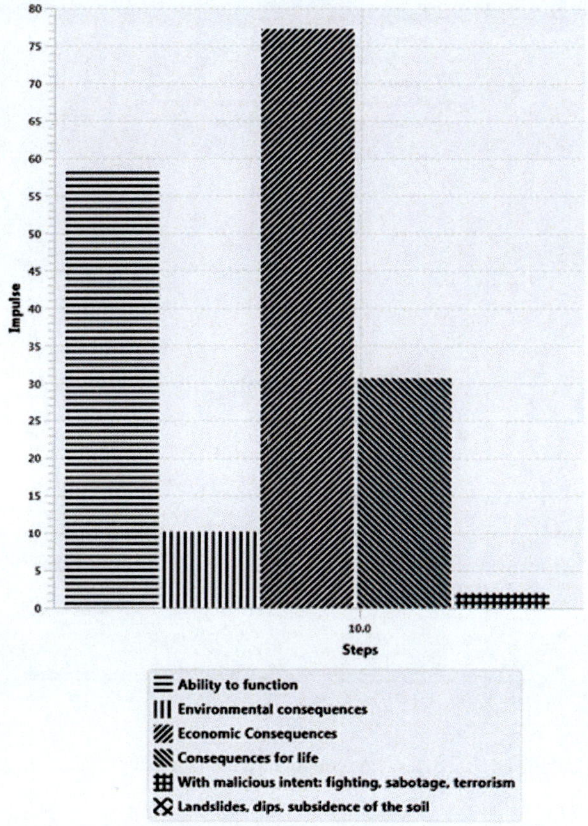

Fig. 6.41 Histograms of pulsed values at the 10th step of modeling. Scenario No. 1

Scenario No. 1. The "pipe duiker" system is affected by anthropogenic factors " With malicious intent: fighting, sabotage, terrorism ". A disturbing impulse $q_{2.1} = +1$ is introduced at the vertex V2-1 (Figs. 6.40 and 6.41).

Scenario No. 2. The system "pipe duiker" is influenced by natural disasters, weather cataclysms (landslides). The disturbing impulse $q_{4-1.1} = +1$ is introduced to the vertex V4–1.1 (Figs. 6.42 and 6.43).

Now consider the similar scenarios for tunnel duikers.

Scenario No. 3. The system "tunnel duiker" is influenced by anthropogenic factors "With malicious intent: fighting, sabotage, terrorism". The disturbing impulse $q_{2.1} = +1$ is introduced to the vertex V2.1 (Figs. 6.44 and 6.45).

Scenario No. 4. The system "tunnel duiker" is influenced by environmental factors (landslides). The disturbing impulse $q_{4-1.1} = +1$ is introduced to the vertex V4.1–1 (Figs. 6.46 and 6.47).

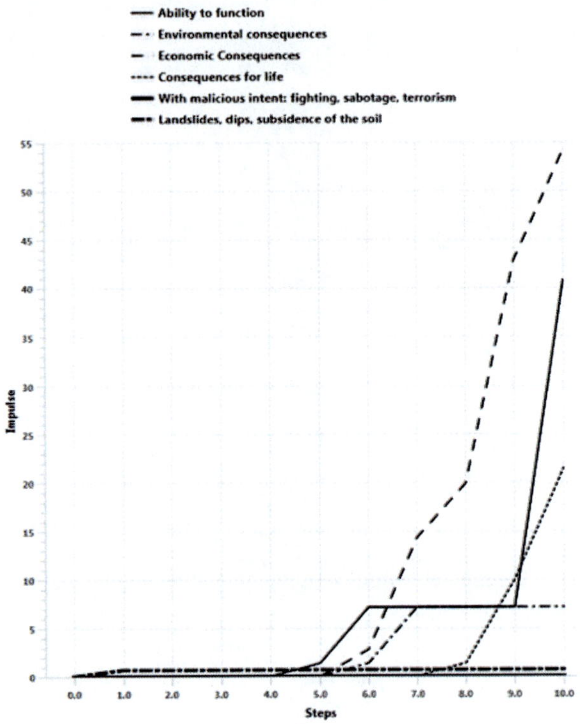

Fig. 6.42 Graphs of pulsed processes, from the first to the tenth step of modeling. Scenario No. 2

The results (see Figs. 6.40, 6.41, 6.42, 6.43, 6.44, 6.45, 6.46 and 6.47) of a numerical experiment of impulse cognitive modeling for pipe and tunnel duikers for anthropogenic events (hostilities, sabotage, terrorism) and natural origin (landslides) are integrated in the form of Table 6.13. The given results of malicious intent impact (military operations, terrorism) (Figs. 6.40 and 6.41; Scenario No. 1) for pipe duikers and for tunnel duikers (Figs. 6.44 and 6.45; Scenario No. 3), show a significant, by 74%, decrease the ability to functioning (V8) and the environmental consequences (V10) of pipe duikers compared to tunnel duikers; by 75% decreasing the consequences of life (V12) and by 71.2% the environmental consequences (V10) of the pipe duikers using. Under the influence of natural origin events (landslides) the ability to functioning (V8) of tunnel duikers (Figs. 5.46 and 5.47; Scenario No. 4) is 57% more reliable than pipe duikers (Figs. 6.42 and 6.43; Scenario No. 2). As follows from the results shown in Table 6.13, the expediency of using tunnel duikers in comparison with pipe duikers for factors environmental consequences (V10), economic consequences (V11), and consequences of life (V12) is 61.5%, 50%, and 62%, respectively.

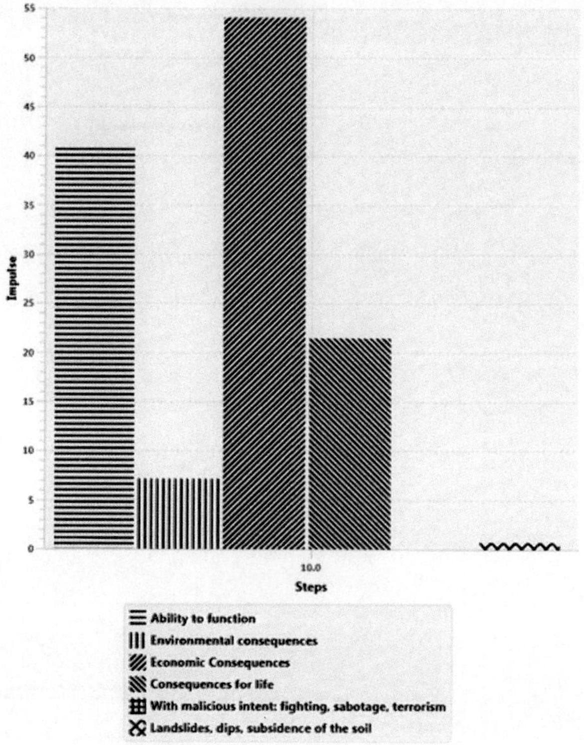

Fig. 6.43 Histograms of pulsed values at the 10th step of modeling. Scenario No. 2

The carried-out studies confirm the priority of the underwater construction of tunnel duikers in comparison with pipe duikers.

Solving this problem is implemented on the basis of the developed strategy of underground construction objects planning based on the foresight and cognitive modeling methodologies. Applying this strategy is especially important for comprehending and preventing negative consequences, minimizing damage under the influence of the most unfavorable combination of negative factors: external and internal static and dynamic loads, all kinds of artificial influences inside an underground structure, harmful natural manifestations from a mountain range, etc.

Combination of the underground planning, geoinformation, experience, equipment, production, and supply of construction materials allows to dramatically reduce costs for construction of underwater tunnel duikers and will increase the quality and safety of people's lives.

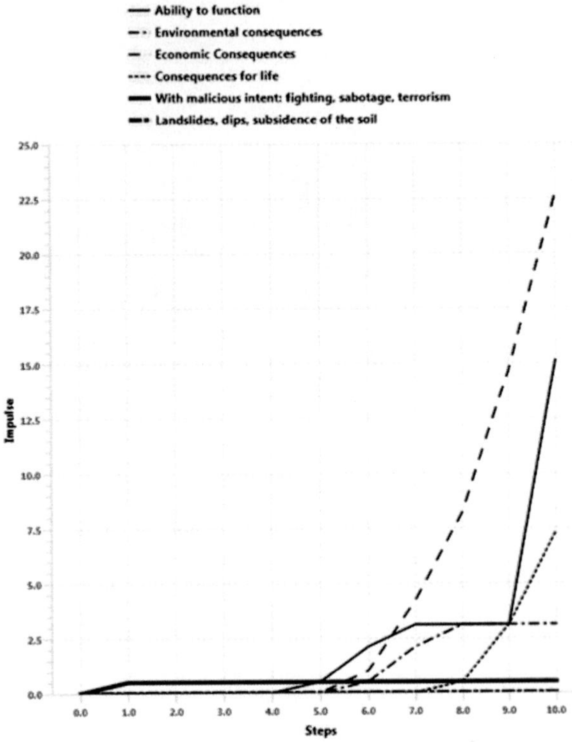

Fig. 6.44 Graphs of pulsed processes, from the first to the tenth step of modeling. Scenario No. 3

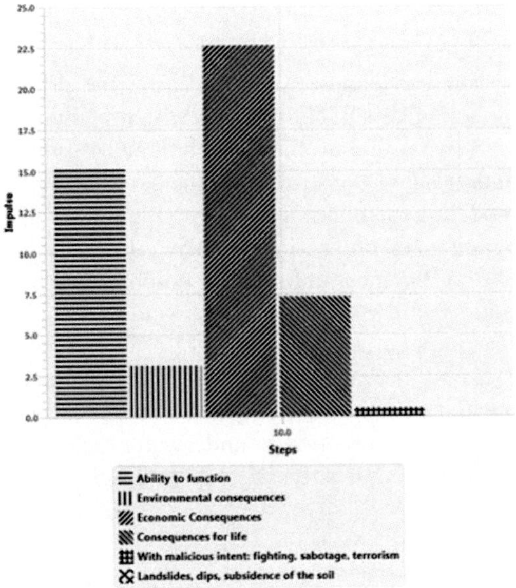

Fig. 6.45 Histograms of pulsed values at the 10th step of modeling. Scenario No. 3

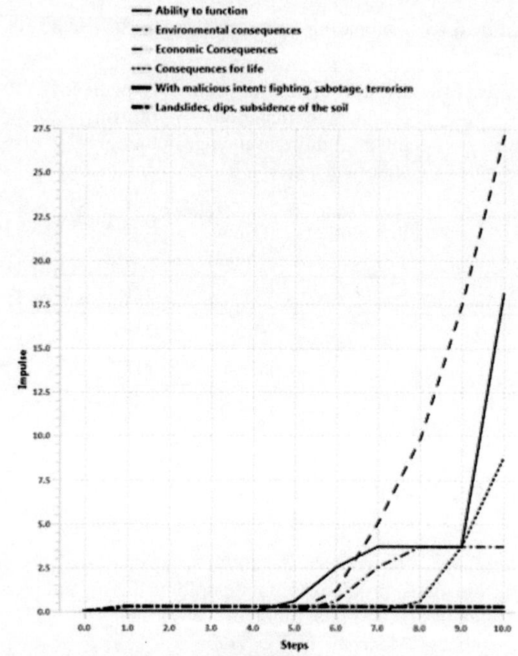

Fig. 6.46 Graphs of pulsed processes, from the first to the tenth step of modeling. Scenario No. 4

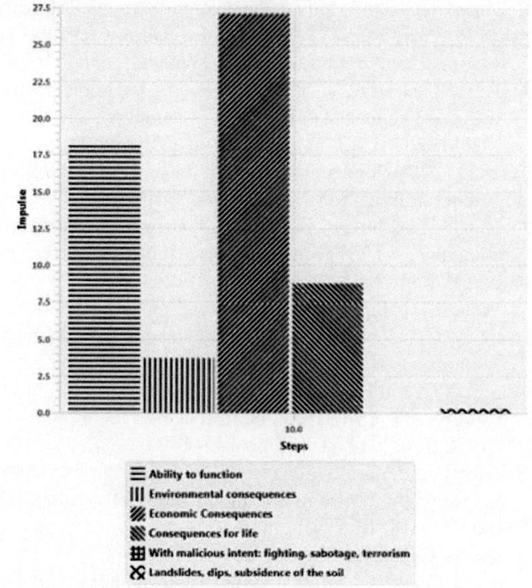

Fig. 6.47 Histograms of pulsed values at the 10th step of modeling. Scenario No. 4

Table 6.13 Comparing values at vertices $V_i (i = 8, 10 - 12)$ for tunnel and pipe duikers at the 10th step of modeling

Vertices	Events of anthropogenic origin. With malicious intent: fighting, sabotage, terrorism $q_{2.1} = +1$		Expediency of using tunnel duikers in relation to pipe duikers	Natural disasters, weather cataclysms: Landslides, dips, subsidence of the soil $q_{4-1.1} = +1$		Expediency of using tunnel duikers in relation to pipe duikers
	Pipe duikers	Tunnel duikers		Pipe duikers	Tunnel duikers	
V8	56	15	74%	41	17,6	57%
V10	10	2,6	74%	7	2,7	61.5%
V11	78	22,5	71.2%	54	27	50%
V12	30	7,5	75%	21	8	62%

References

Abramova N, Avdeeva Z (2008) Cognitive analysis and management of the development of situations: problems of methodology, theory and practice. Probl Control 3:85–87

Alptekin N (2013) Integration of SWOT analysis and TOPSIS method in strategic decision making process. Macrotheme Rev 2(7)

Atkin R (1997) Combinatorial connectivies in social systems. An application of simplicial complex structures to the study of large organisations. Interdisciplinary Systems Research

Atkin R, Casti J (1977) Polyhedral dynamics and the geometry of systems. RR-77-International Institute for Applied Systems Analysis, Laxenburg, Austria

Avdeeva Z, Kovriga S (2018) On governance decision support in the area of political stability using cognitive maps. In: 18th IFAC conference on technology, culture and international stability (TECIS2018), vol 51(30), pp 498–503

Casti J (1979) Connectivity, complexity, and catastrophe in large-scale systems. A Wiley—Interscience Publication International Institute for Applied Systems Analysis. JOHN WILEY and SONS. Chichester – New York – Brisbane – Toronto

Gantmakher F (1967) Matrix theory. Nauka, Moscow (In Russian)

Gilbert P et al (2013) Underground engineering for sustainable urban development. The National Academies Press, Washington

Ginis L, Gorelova G, Kolodenkova A (2016) Cognitive modeling of development of regional economy system. Int J Econ Finan Issues 6(5):97–103

Gorelova G, Pankratova N (2018) Scientific foresight and cognitive modeling of socio-economic systems. In: 18-th IFAC conference on technology, culture and international stability, TECIS-2018, IFAC, vol 51(30), pp 145–149

Gorelova G, Pankratova N (eds) (2015) Innovative development of socio-economic systems based on foresight and cognitive modelling methodologies. Naukova Dumka, Kyiv (In Russian)

Haiko H (2018) Mastering underground space in the concept of sustainable development of large cities. Geotechnologies 1:60–64 (In Ukrainian)

Kairong H (2017) Typical underwater tunnels in the mainland of china and related tunneling technologies. Engineering 3(6):871–879

Kartosiya B (2015) Mastering underground space of large cities. New Trends. Sci Inf Anal Bull (sci Tech j) 1:615–629 (In Russian)

Konykhov S (2010) Systematization of approaches for developing underground city space. Vestnik MGSU 4:56–61

Kovriga S, Maksimov V (2001) Cognitive technology of strategic management of the development of complex socio-economic objects in an unstable external environment. 1st issue of Cognitive Analysis and Situational Management (CASC'2001)

Kulba V, Kononov D, Kovalevsky S, Kosyachenko S, Nizhegorodtsev R, Chernov I (2002) Scenario analysis of the dynamics of behavior of socio-economic systems. IPU RAS, Moscow

Kulikova E, Korchak A, Levchenko A (2005) Strategy of risk management in urban underground construction. Moscow State University State University Publishing House, Moscow (In Russian)

Levchenko A (2007) About a new direction of scientific research in construction geotechnology. Min Inf Anal Bull (sci Tech j) 2:15–21

Maksimov V (2001) Cognitive technology—from ignorance to understanding. 1st work "Cognitive analysis and management of the development of situations", (CASC'2001), pp 4–18

Mardani A, Zavadskas E, Govindan K, Senin A, Jusoh A (2016) VIKOR technique: a systematic review of the state of the art. Literat Methodol Appl Sustain 8:1–38

Mikhnenko P (2015) Dynamic modification of SWOT analysis. Econ Anal Theory Pract 18(417):60–68

Owen C, Bezerra C (2000) Evolutionary structured planning. A computer-supported methodology for the conceptual planning proces. In: Gero JS (ed) Artificial intelligence in design'00. Kluwer Academic Publishers, Dordrecht, pp 287–307

Pankratov V (2014) Development of the approach to formalization of vector's indicators of sustainable development. J Inf Technol Knowl 8(3):203–211. ITHEA, Sofia

Pankratova N, Gayko G, Kravets V, Savchenko I (2016) Problems of megapolises underground space system planning. J Autom Inf Sci 48(4):32–38

Pankratova ND, Gorelova GV, Pankratov VA (2020) Study of the plot suitability for underground construction: cognitive modelling. In: ISDMCI 2020: Lecture Notes in Computational Intelligence and Decision Making vol 1246, pp 246–264. https://doi.org/10.1007/978-3-030-54215-3_16

Pankratova N, Malafieieva L (2017) Delphi method. Theory and applications. Reference Book. Naukova Dumka, Kyiv. (In Ukrainian)

Pankratova, N., Nedashkivska, N.: Models and Methods of Hierarchy Analysis. Theory. Applications: Reference Book. "Polytechnica" Publishing, Kyiv (2010). (In Ukrainian)

Pankratova ND, Pankratov VD (2022) System approach to the underground construction objects planning based on foresight and cognitive modelling methodologies. Syst Res Inf Technol 1:7–25. https://doi.org/10.20535/SRIT.2308-8893.2022.1.01

Pankratova N, Savastiyanov V (2014) Foresight process based on text analytics. Int J Inf Content Process 1(1): ITHEA, Sofia

Pankratova N, Savchenko I (2015) Morphological analysis. Problems, theory, application. Naukova Dumka, Kyiv

Pankratova N, Savchenko I, Haiko H, Kravets V (2019a) System approach to planning urban underground development. Int J Inf Content Process 6(1):3–17

Pankratova N, Gorelova G, Pankratov V (2019b) Strategy for the study of interregional economic and social exchange based on foresight and cognitive modeling methodologies. In: Proceedings of the 8th international conference on mathematics. Information technologies. Education, pp 136–141, Shatsk, Ukraine

Pankratova N, Gorelova G, Pankratov V (2022) The strategy of underground construction objects planning based on foresight and cognitive modelling methodologies. In: Zgurovsky M, Pankratova N (eds) System analysis & intelligent computing, studies in computational intelligence, vol 1022, pp 69–91. https://doi.org/10.1007/978-3-030-94910-5_5

Roberts F (1978) Graph theory and its applications to problems of society. Society for Industrial and Applied Mathematics, Philadelphia

Sakellariou M (2020) Tunnel engineering—selected topics. National Technical University of Athens, Books, IntechOpen 6201, Athens

Saługa P (2009) Ocena ekonomiczna projektów i analiza ryzyka w górnictwie [Economic Evaluation and Risk Analysis of Mineral Projects]. Studia, Rozprawy, Monografie, nr 152, Wyd. IGSMiE PAN, Kraków

Sterling R, Admiraal H, Bobylev N, Parker H, Godard J, Vähäaho I, Shi X, Hanamura T (2012) Sustainability issues for underground spaces in urban areas. Proc. ICE. Urban Des. Plan. 165(4):241–254

Vernadsky V (2012) Biosphere and noosphere. Iris Press, Moscow (In Russian)

Weimer-Jehle W (2006) Cross-impact balances: a system-theoretical approach to cross-impact. Technol Forecast Soc Chang 73:334–361

Zgurovsky M, Pankratova N (2005) Technology Foresight. "Polytechnica" Publishing, Kyiv. (In Ukrainian)

Zgurovsky M, Pankratova N (2007) System analysis: theory and applications. Springer

Zgurovsky M, Pankratov V (2014) Strategy of innovative development of the region based on the synthesis of foresight methodology and cognitive modelling. Syst Res Inf Technol 2:7–17